04-DEC-7

RESEARCH DIGEST

FACTORS AFFECTING THE PERFORMANCE OF PRIMARY TREATMENT IN DECENTRALIZED WASTEWATER SYSTEMS

by:

Victor A. D'Amato, PE
ARCADIS U.S., Inc.

Anita Bahe, Ph.D.
Lynx Group International

Terry Bounds, PE
Orenco Systems, Inc.

Billie Comstock
ARCADIS U.S., Inc.

Thomas Konsler
Orange County (NC) Environmental Health

Sarah K. Liehr, PE, Ph.D.

Sharon C. Long, Ph.D.
WSLH/University of Wisconsin

Krich Ratanaphruks
Wayla Solutions, LLC

Chet A. Rock, PE, Ph.D.
University of Maine - Orono

Kevin Sherman, PE, Ph.D.
Quanics, Inc.

2008

WERF

The Water Environment Research Foundation, a not-for-profit organization, funds and manages water quality research for its subscribers through a diverse public-private partnership between municipal utilities, corporations, academia, industry, and the federal government. WERF subscribers include municipal and regional water and wastewater utilities, industrial corporations, environmental engineering firms, and others that share a commitment to cost-effective water quality solutions. WERF is dedicated to advancing science and technology addressing water quality issues as they impact water resources, the atmosphere, the lands, and quality of life.

For more information, contact:
Water Environment Research Foundation
635 Slaters Lane, Suite 300
Alexandria, VA 22314-1177
Tel: (703) 684-2470
Fax: (703) 299-0742
www.werf.org
werf@werf.org

This report was co-published by the following organizations. For non-subscriber sales information, contact:

IWA Publishing
Alliance House, 12 Caxton Street
London SW1H 0QS, United Kingdom
Tel: +44 (0) 20 7654 5500
Fax: +44 (0) 20 7654 5555
www.iwapublishing.com
publications@iwap.co.uk

Library of Congress Catalog Card Number: 2007920273
IWAP ISBN: 978-1-84339-781-6

ARCADIS U.S., Inc.

The research on which this report is based was developed, in part, by the United States Environmental Protection Agency (EPA) through Cooperative Agreement No. X-830851-01 with the Water Environment Research Foundation (WERF). However, the views expressed in this document are solely those of ARCADIS U.S., Inc. and neither EPA nor WERF endorses any products or commercial services mentioned in this publication. This report is a publication of WERF, not EPA. Funds awarded under the Cooperative Agreement cited above were not used for editorial services, reproduction, printing, or distribution.

ACKNOWLEDGMENTS

Thanks to the Project Team, WERF Project Managers and Project Subcommittee, and to the Project Team for WERF Project No. 04-DEC-1 "Influent Constituent Characteristics of the Modern Waste Stream from Single Sources" (Dr. Kathryn Lowe, PI) for continued support and feedback during this project. A special thanks is extended to others outside of the project who offered their interest, insight and assistance, including, Jim Kreissl, Barbara Rich, Craig Jowett, Peter Gavins, Peggy Minnis, Steven Berkowitz, Barbara Grimes, Carl Etnier, Joel Ducoste, Joe Harrison, the staffs of the National Small Flows Clearinghouse, National Onsite Wastewater Recycling Association (NOWRA), and many others. Their selfless devotion to the decentralized wastewater management field is inspiring and their contribution to this project is appreciated.

Report Preparation

Principal Investigator:
Victor A. D'Amato, PE
ARCADIS U.S., Inc.

Project Team:
Anita Bahe, Ph.D.
Lynx Group International

Terry Bounds, PE
Orenco Systems, Inc.

Thomas Konsler
Orange County (NC) Environmental Health

Sarah K. Liehr, PE, Ph.D.

Sharon C. Long, Ph.D.
WSLH/University of Wisconsin

Krich Ratanaphruks
Wayla Solutions, LLC

Chet A. Rock, Ph.D., PE
University of Maine - Orono

Robert Seabloom, PE
University of Washington

Kevin Sherman, Ph.D., PE
Quanics, Inc.

Billie Comstock
ARCADIS U.S., Inc.

Project Subcommittee

Matt Byers, Ph.D.
Zoeller Company

Bruce Lesikar, Ph.D.
Texas A&M University

Richard J. Otis, Ph.D., PE, DEE
Ayres Associates.

Kyoung S. Ro, Ph.D., PE
United States Department of Agriculture, Agricultural Research Service, Coastal Plains Soil, Water, & Plant Research Center

Saya Qualls, PE
Tennessee Department of Environment & Conservation

Water Environment Research Foundation Staff

Director of Research: Daniel M. Woltering, Ph.D.
Program Manager: Christine Handog

DEDICATION

Dedicated to the memory of Robert (Bob) W. Seabloom (June 18, 1924 - February 12, 2007), friend and colleague, and a national leader in the study and teaching of decentralized wastewater treatment. Bob spent his entire working life teaching Civil Engineering at the University of Washington and always loved his Huskies. In his personal life, he enjoyed traveling, playing handball and tennis, skiing and first and foremost, spending time with his family and many friends. Bob was a great and enthusiastic individual, and his dedication and efforts to bring knowledge and understanding to all of us will be deeply missed.

ABSTRACT AND BENEFITS

Abstract:

Approximately 23% of the estimated 115 million occupied homes in the United States are served by decentralized wastewater systems, the vast majority of which include a septic tank, grease trap, or both for primary treatment. These are efficient, simple treatment units whose performance is critically important to the overall functioning of decentralized wastewater systems.

Current primary unit design and operational practices tend to be driven by highly prescriptive regulations, industry standards and guidance materials which rely on limited and often dated scientific information and past practices and may be incomplete in their consideration of the factors that may influence primary treatment unit performance in decentralized wastewater systems. Although septic tanks and grease traps are generally robust and efficient primary treatment units, the optimization of their design and operation has been limited. Further research may be warranted to answer outstanding questions and optimize practices; however it should be conducted within the framework of overall decentralized system performance objectives and functions.

The objective of this project was to take a fresh look at the existing body of work addressing the performance of primary treatment units in decentralized wastewater systems, with a goal of establishing what is known, what is not known and what future research may be warranted. Design, construction/installation, and operation, monitoring and maintenance issues were each considered, with a focus on those factors most likely to affect primary unit treatment objectives, including influent characteristics, sizing, hydraulic design, compartmentation, influent and effluent appurtenances and seasonal effects, among others.

Relevant publications were thus identified, compiled, analyzed and synthesized with important research recommendations defined to support the development of several interrelated products, designed to be useful tools for WERF subscribers, practitioners, researchers, policy-makers and other stakeholders:

- Research Digest
- Bibliographic Database
- (3) Communications Documents (Policy, Research, Technical)
- User-friendly compact disk—read only memory (CD-ROM) for web interface

Benefits:

- Compiles information on what is known about the performance of primary treatment in decentralized wastewater treatment systems and factors affecting performance
- Describes the landscape for primary treatment in decentralized wastewater treatment systems: history, regulations, industry standards and standard practice
- Sets basis and provides recommendations for future research on primary treatment in decentralized wastewater systems
- Provides bibliographic database to assist researchers and practitioners

Keywords: Septic tank, grease trap, decentralized wastewater, anaerobic reactor.

TABLE OF CONTENTS

LIST OF TABLES

LIST OF FIGURES

LIST OF ACRONYMS

AF	Anaerobic Filters
ASTM	American Society for Testing and Materials
AWWA	American Water Works Association
BAST	Baffled Anaerobic Septic Tanks
BASTAF	Baffled Anaerobic Septic Tanks with Anaerobic Filter
BOD	Biochemical Oxygen Demand
BOD_5	Biochemical Oxygen Demand, 5-day
BOD_7	Biochemical Oxygen Demand, 7-day
cBOD	Carbonaceous Biochemical Oxygen Demand
CD-ROM	Compact Disk - Read Only Memory
CEN	European Committee for Standardisation
CFD	Cumulative Frequency Diagrams
CN	Cyanide
COD	Chemical Oxygen Demand
CSA	Canadian Standards Association
DW	Drinking Water
EPA	US Environmental Protection Agency
FC	Fecal Coliform
FOG	Fats, Oils and Grease
HHFA	Housing and Home Finance Agency
HRT	Hydraulic Retention Time
IAPMO	International Association of Plumbing and Mechanical Officials
ICC	International Code Council
IPC	International Plumbing Code
MCL	Maximum Contaminant Level
NDWRCDP	National Decentralized Water Resources Capacity Development Project
NOWRA	National Onsite Wastewater Recycling Association
NPCA	National Precast Concrete Association
NSF	National Sanitation Foundation
O&G	Oil and Grease
O&M	Operation and Maintenance
ORP	Oxidation Reduction Potential

PHS	Public Health Service
QA	Quality Assurance
QC	Quality Control
RV	Recreational Vehicle
SAR	Sodium Absorption Ratios
SCL	Secondary Contaminant Level
SLR	Surface Loading Rate
SS	Suspended Solids
STE	Septic Tank Effluent
STEG	Septic Tank Effluent Gravity
STEP	Septic Tank Effluent Pump
TP	Total Phosphorous
TS-1	Primary Treatment Standard
TS	Total Solids
TSS	Total Suspended Solids
UASB	Upflow Anaerobic Sludge Blanket
UL	Underwriters Laboratories
UPC	Uniform Plumbing Code
V/Q	Volume/Flow Rate
VFA	Volatile Fatty Acids
VOA	Volatile Organic Acids
VOC	Volatile Organic Compounds
WQA	Water Quality Association

EXECUTIVE SUMMARY

Approximately 23% of the estimated 115 million occupied homes in the United States are served by decentralized wastewater systems. This proportion has not changed significantly for at least 30 years and—due in part to exurban development trends as well as the country's infrastructure funding challenges—is not expected to change much in the next 30 years or more. The vast majority of decentralized wastewater treatment systems include a septic tank, grease trap, or both for primary treatment. These primary treatment units may precede non-traditional collection systems, additional secondary or advanced external treatment units such as biological filters, constructed wetlands, aerobic treatment units, and disinfection systems, or they may directly precede discharge to a soil-based treatment and disposal system. Secondary or advanced treatment systems, and especially soil-based treatment systems, can be quite sensitive to the quality of the primary effluent they receive. The performance of the primary units in a decentralized wastewater system is therefore absolutely critical, though frequently overlooked due to the relative simplicity of these units and the tendency to focus attention on more glamorous aspects of the decentralized wastewater system design. Downstream problems in decentralized wastewater treatment systems are frequently traced back to the performance of the primary treatment units.

While primary treatment units in decentralized wastewater systems are of relatively simple design and operation, septic tank function is actually quite complex, performing primary settling, sludge storage and anaerobic digestion in a single reactor. Septic tanks are horizontal flow reactors in which primary setting occurs and aerobic, anaerobic and facultative organisms perform complex biochemical processes which may take two or more years to mature (Seabloom et al., 2005). Grease traps are expected to cool incoming food service wastewater for gravity separation and storage of fats, oils and grease (FOG), via physical and physiochemical processes.

The design, construction, installation, and maintenance of primary treatment units in the U.S. are driven primarily by prescriptive state and local regulatory standards, and secondarily by guidance manuals/engineering texts and industry standards. Some of the most influential research dates back to US Public Health Service [US PHS, a predecessor to the US Environmental Protection Agency (EPA)] studies in the mid-20th century (Weibel et al., 1949, 1954, 1955). Many of the results of the US PHS work are reflected in modern septic tank design and construction practice and reliance on the septic tank designs evolving from this work has restricted the continued development of design improvements, except in relatively simple enhancements to key structures. Meander tanks, closed-conduit, laminar flow tanks, upflow anaerobic sludge blanket (UASB)-septic tanks and Imhoff tanks are alternatives to conventional septic tank designs but currently are of only limited, international, or niche use.

Performance Factors

Treatment Efficacy

Septic tanks (and grease traps) primarily remove solids (mostly the settleable fraction) and FOG from influent wastewater. Organic carbon is cycled through settling to the sludge layer and subsequent resolublization. A portion of the influent organic carbon is biologically transformed in septic tanks, partially as a result of biological activity in the supernatant or clear zone. Nitrogen is typically transformed from organic nitrogen in the influent to ammonia forms. The fate of other influent constituents is primarily controlled by physiochemical processes, including sorption and settling.

Facilities

Most work to optimize septic tank design has been based on studies on single residential sources (household wastewaters). Septic tank performance has been studied in detail for few other facility types and in fact, influent wastewater quality characteristics have been reported for relatively few facility types. Residential clusters, vehicle service stations, recreational areas, highway rest areas and funeral homes have been studied and reported; however, the potential use of the published information to inform primary unit design is limited.

Sizing

Septic tank sizing is typically based on allocating volume for sludge storage, clarified supernatant, a floating scum layer and air space for surge storage and venting. Hydraulic residence times of one to three days are common and sizing is at least as important with regards to pumping frequency as it is in terms of settling efficiency, with larger tanks taking longer to reach biological maturity, but also ultimately being more efficient digesters in addition to having larger capacities for sludge storage.

Hydraulic and Settling Efficiency

Hydraulic surges can be managed by maximizing surface area and restricting intercompartment transfer and effluent flowrate. Increased surface area also improves the settling efficiency of septic tanks. Settling efficiency, however, may be adversely affected by the resuspension of solids as a result of biogas ebullition from the digesting sludge layer. As such, compartmentation (which may also improve hydraulic performance) and the use of effluent screening devices can mitigate the negative impacts of sludge digestion. Hydraulic efficiencies can also be improved by the use of tanks with long, relatively narrow aspects and by considering, especially, inlet flow dispersion. Data which isolates the effects of specific dimensional criteria as well as inlet, intercompartment transfer and outlet design are lacking.

Food Service Facilities and Grease Traps

Very little is known about the factors impacting the performance of grease traps and likewise it appears that almost nothing has been done to optimize design or to confirm that sizing criteria in use today (which varies widely) is appropriate. What is known is that large outdoor grease traps similar in design to septic tanks are necessary to remove substantial amounts of grease in decentralized wastewater treatment systems treating food service facilities. Additionally, it is clear that management and operating practices of restaurants are critically important grease control measures. Practitioners agree that blackwater should only be co-mingled downflow of the grease and oil separation units, so that removal of accumulated materials for rendering is viable.

Seasonal Effects

The well-understood dependence of biological reaction rates on temperature has been clearly observed in laboratory and field observations of operating septic tanks. Numerous researchers have anecdotally described a "spring turnover" or "boil" and often an increase in effluent solids concentrations during warmer months. Settling and solids removal efficacy increases during the cooler months coincident with a growth in the amount of accumulated solids. In the warmer months, digestion and solids reduction is maximized, reducing the amount of accumulated solids, but gas ebullition during the increased digestion may hinder settling and limit solids removal efficiency.

Solids Accumulation and Removal

Sludge and scum accumulation rates have been established by a number of researchers over the years with results in fairly good agreement with one another. Although a number of factors will affect the rate of solids accumulation, these relationships can be used as a general guide for designers and planners. The consistent theme from these investigations is that the typically-recommended pump-out interval of 3-5 years is often quite conservative and, more importantly, to the possible detriment of the system since methanogenesis does not fully develop until after about three years of operation. While this conservatism may be warranted, as it acknowledges the historical lack of oversight and maintenance afforded to decentralized wastewater systems, accurate, periodic measurement of accumulated materials is the only way to truly determine pump-out needs.

Operation and Monitoring

The bulk of the information on additives suggests that bioaugmentation products are generally benign but of marginal benefit. Chemical additives should be avoided due to their potential for harming tank biology and contaminating the receiving environment. Information on the impacts of water softener regenerant brine are inconclusive and in need of further evaluation. Protocols for performance monitoring and troubleshooting are quite important. Remote monitoring tools are gaining in popularity as are methodologies for evaluating septic tank biology to aid in troubleshooting performance problems. The impacts of household chemicals on septic tank biology can be assessed via screening tests.

Construction and Installation

Despite the limited amount of published information regarding septic tank construction and installation practices, most field practitioners agree that tank construction and installation are critical and that improper construction and/or installation has been responsible for many system malfunctions. As such, an emphasis on this element of tank "performance" is warranted. Construction and installation practices for tanks of various materials installed under various conditions are generally known; that is, fundamental research is not likely to be warranted. However, the further development of procedures for ensuring proper construction and installation – quality assurance/quality control (QA/QC)—particularly in the area of watertightness is an area needing work.

Research Recommendations

There are a multitude of potential research questions that could be pursued. However, the answers to many such questions are likely to be inconsequential; that is, the results will not result in meaningful improvements in system performance. Furthermore, the resources required to answer some questions with confidence may not be worth the potential improvements in practice. There is thus a great need to consider all potential research objectives (20 of which are categorized in the body of this digest) with respect to anticipated research costs versus potential benefits to overall decentralized system performance.

Future studies should carefully consider the applicability of the three main types of studies that are represented in the existing body of research: laboratory-scale studies, controlled pilot testing and field surveys. Each is appropriate in certain circumstances and of limited utility in others.

Additionally, common study pitfalls should be avoided. These include:

- Conducting studies with too many variables, making it impossible to statistically isolate the effects of individual variables on performance and thus yielding little useful information. Future studies need to be more realistic and set clear research objectives that can be accomplished with available resources. Additionally, such well-designed research should be published in respected, peer-reviewed journals; there is currently a paucity of such publications in the literature.
- Conducting controlled pilot studies by adding sewage sludge (often from a municipal wastewater treatment plant) to a septic tank to simulate natural septage accumulation. This practice is unrealistic in its consideration of the important biological functions and effects inherent to septic tanks.
- Heavy reliance on septic tank effluent (STE) data as the primary performance measure. Because influent characteristics are such a fundamental variable inherent to virtually all controlled pilot and field survey experimental designs, paired influent and effluent data may be needed to adequately characterize performance.

During project planning, it was assumed that there was an existing untapped pool of data that could be used to inform primary unit practice in a meaningful way. However, it appears that most of the truly useful data in the field has been published in some forum. What information that is available in the decentralized wastewater arena could be better managed and distributed. The industry as a whole should attempt to develop tools that will allow future (and perhaps existing) monitoring data to be banked collectively.

Future studies can use the results of this project to identify useful existing data; however, it is recommended that the focus be on using available resources to conduct original research, focusing on statistically robust experimental designs.

Additional Information

Several interrelated products associated with the research digest are available. The following products are available at www.werf.org and www.ndwrcdp.org:

- **Research Digest**: Factors Affecting the Performance of Primary Treatment in Decentralized Wastewater Systems
  - Research Digest Appendices
- **Bibliographic Database** of Research and Data on Performance of Primary Treatment Units in Decentralized Wastewater Systems
  - Bibliographic Database Readme
- **Fact Sheets and Communications Tools**
  - Technical Guide: Primary Treatment in Decentralized Wastewater Systems (succinct guidance document)
  - Primary Treatment in Decentralized Wastewater Systems: Research Considerations (fact sheet)
  - Primary Treatment in Decentralized Wastewater Systems: Policy Points (fact sheet)

Background

Primary Unit Objectives

The objectives of primary units in decentralized wastewater treatment systems may vary somewhat depending on factors such as facility/wastewater type, downstream system component and receiving environment sensitivities, and others. That said, septic tanks serving traditional sources are generally expected to:

1. Separate settleable solids as sludge and floatable materials as part of the scum layer.
2. Temporarily store separated materials.
3. Digest accumulated materials. Note that this digestion has both positive and negative consequences, as indicated in Table 1. On balance, the function would generally be considered positive.

Table 1. Positive and Negative Effects of Digestion in Septic Tanks.

Positive	Negative
♦ Volume reduction, less frequent pumping ♦ Volume reduction, increased effective volume/settling capacity ♦ Production of potentially-recoverable biogas ♦ Phase conversion of organics from solid to dissolved (liquid) ♦ Ebullition of gas bubbles, "seeding" clear and scum zones with active biomass	♦ Ebullition of gas bubbles, resuspending solids in clear zone and effluent ♦ Phase conversion of organics to dissolved phase may increase soluble biochemical oxygen demand (BOD) over influent levels ♦ Ebullition of gas bubbles, helping build a floating scum mat (the positive and negative effects of this scum layer are debatable)

The overriding treatment objective is to produce a consistent, conditioned, clarified effluent suitable for downstream processing (often via a soil absorption system), with relatively infrequent removal of accumulated materials.

Grease traps have functions and objectives that are somewhat different than those of septic tanks. Whereas septic tanks are clearly anaerobic digesters as well as primary sedimentation chambers, grease traps are expected to function almost entirely as physical separation units (though, undoubtedly biological activity occurs and may, or may not, affect performance). Thus, the functions of grease traps are to:

1. Cool influent wastewater for gravity separation of fats, oils and grease (FOG).
2. Store accumulated FOG, generally as a floating layer.

Like septic tanks, the overall objective of grease trap treatment is to reduce challenges to downstream system components. The removal of accumulated material in a grease trap is generally much more frequent than for septic tanks. As for septic tanks, the removal frequency is dependent on the size and storage capacity of the unit. Unlike septic tanks, there is generally no functional performance advantage to storing the grease, so removal is typically done on a regular rather than as-needed basis.

Biological Model of Septic Tank Function

Septic tanks are horizontal flow reactors in which primary setting occurs and aerobic, anaerobic and facultative organisms perform complex biochemical processes (Seabloom et al., 2005). A process schematic of a properly operating septic tank is illustrated in Figure 1. Anaerobic processes dominate due to the lack of oxygen typically in the liquor, although oxygen-rich influent slugs may periodically influence the biology, particularly in the clear zone.

In the first stage of biological development in a septic tank – the acid or non-methanogenic phase—facultative acid-forming bacteria hydrolyze complex organic molecules into simple soluble organic compounds (liquefaction), including simple sugars (for starches) and amino acids (for proteins). Fats remain essentially intact. Continued metabolism generates organic acids accompanied by a reduction in pH and potential inhibition of further bacterial decomposition. A second group of bacteria—the methane formers—metabolize the organic acids to carbon dioxide and methane. Breakdown of the amino acids liberates ammonia which has a tendency to raise the pH to a favorable level for methanogenesis (Seabloom et al., 2005). However, temperatures may be somewhat restrictive to full methanogenesis, thus allowing volatile fatty acids (VFA) to pass in the effluent from the septic tank, measurable as biochemical oxygen demand (BOD). The methanogens, which are strict anaerobes, are also able to attack fats, generating simpler compounds that can be further metabolized to methane and carbon dioxide.

Anaerobic digestion in the settled sludge layer generates gas bubbles that potentially carry microorganisms as they rise, seeding the liquid zone between the sludge and scum layers and facilitating decomposition of the soluble organic material in the bulk liquid (Baumann and Babbitt, 1953). In properly operating septic tanks, three distinct vertically oriented layers should form: a floating scum layer, a relatively clear layer in the middle, and a lower settled sludge layer. The clear zone is generally anoxic, containing bound oxygen, with moderate oxidation reduction potential (ORP), while the sludge and scum layers are more strictly anaerobic, containing no free or bound oxygen and exhibiting a highly negative ORP. The clear zone is anoxic due in part to the introduction of oxygen-rich influent into that zone. It may take several years to develop volatile organic acids (VOA) and metabolite concentrations suitable for a stable methanogenic population and optimum digestion, depending on the design of the tank and the characteristics of the wastewater feed (Bounds, 1997b).

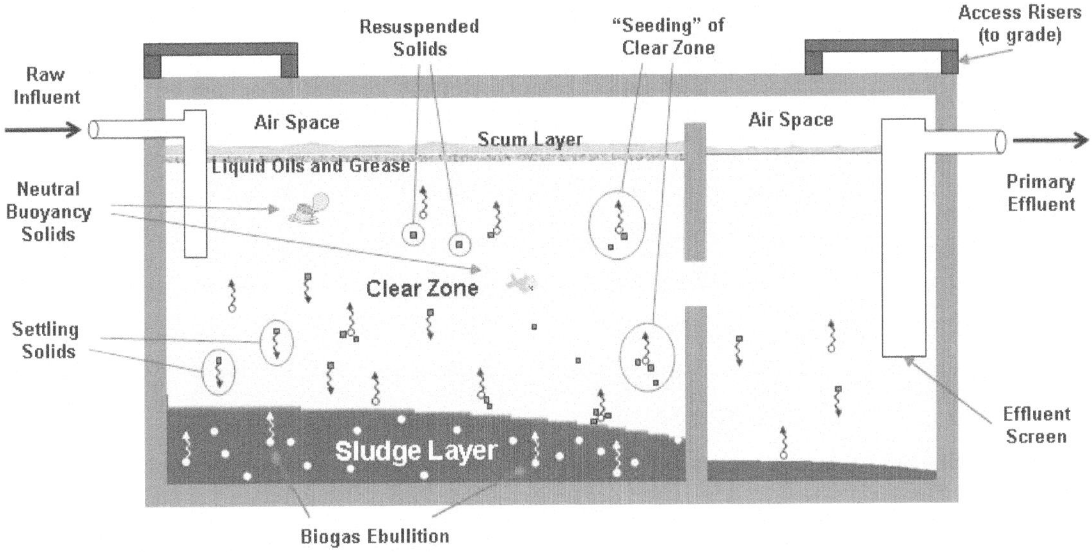

Figure 1. Process Schematic of Properly Operating Septic Tank.

Modern History and Evolution of Primary Treatment Units

Modern septic tank design has evolved mostly as a function of construction convenience, low cost and repetitive practice (Winneberger, 1984). In the mid-century, beginning around 1946 and continuing into 1953, the U.S. federal government conducted an extensive and influential testing program for decentralized technologies and specifically, septic tanks. The work was initiated by the Federal Housing and Home Finance Agency (HHFA) and conducted by the U.S. Public Health Service (U.S. PHS), and resulted in three major reports (Weibel et al., 1949, 1954, 1955).

Many of the results of the U.S. PHS work are reflected in modern septic tank design and construction practice. However, a fundamental limitation of the PHS work was that the starting point for testing was septic tank designs that were already in use, rather than potentially enhanced designs. It is understandable that the PHS work took this pragmatic approach, as a primary driver for their work was the high failure rates of existing in-service systems and the liabilities imposed upon the federal government (e.g., homeowner defaults on federally-subsidized loans). Despite its several limitations—which also included a reliance on controlled pilot-testing conditions over field surveying and the extensive use of added sewage sludge as a surrogate for accumulated septage—the U.S. PHS studies yielded septic tank designs that were fairly robust within the constraints of the testing that was conducted. An example schematic showing a typical two-compartment septic tank, generally consistent with the results of the U.S. PHS research and subsequent evolution of design is provided in Figure 2.

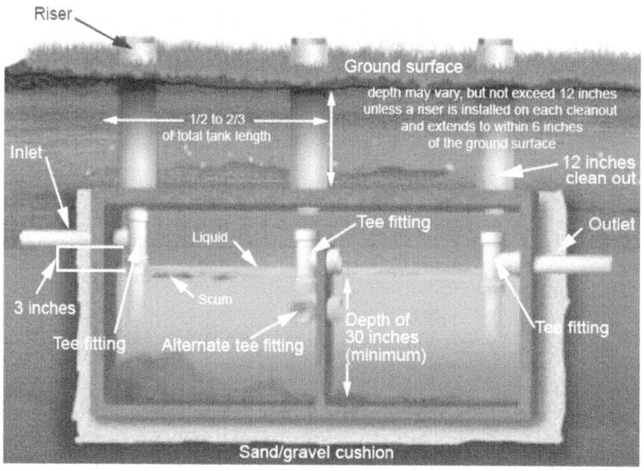

Figure 2. Profile Schematic of a Typical Two-Compartment Septic Tank (courtesy of Texas A&M University).

Unfortunately, reliance on the septic tank designs evolving from the PHS work—and the codification of these designs into prescriptive state regulations—has served to restrict the continued development of design improvements, except in relatively simple enhancements to key structures (effluent screening being a relevant example). More radical, paradigm-shifting designs have been difficult to develop commercially, as they may be more expensive to construct (especially at first) and may require significant deviations from prescriptive standards and regulatory codes.

A thorough description of the current landscape of regulations and industry standards can be found in Appendix A and a more complete history of septic tanks can be found in Appendix B.

Sedimentation Theory and its Relevance to Tank Design

Ideal settling theory has been used to describe the settling behavior of particles in centralized wastewater treatment processes for both primary and secondary clarification. In conventional primary and secondary clarification, the treatment process is predominantly physical in nature and accordingly the design generally assumes that biological processes are negligible. This is a reasonable assumption, since solids are removed from these units - via wasting or recycling - on a regular or semi-continuous basis, particularly compared to septic tanks, and the hydraulic and solids residence times in these units are accordingly relatively low. In fact, anaerobic digestion and gas ebullition in secondary clarifiers is typically identified as an operational problem that adversely affects settling. Septic tanks, with their dual function as settling tanks and digesters, are different, requiring a more nuanced application of ideal settling theory to assessments of their performance and design.

Ideal settling theory and its application in water/wastewater engineering is attributed to Hazen (1904) and has subsequently been developed and applied and reported in many wastewater engineering texts and guidance materials. Four types of settling phenomena are described and each is likely to be occurring to some extent in properly operating septic tanks. The processes are as described in Table 2 and depicted in Figure 2 (Seabloom et al., 2005, following Camp, 1946).

Table 2. Description of Settling Phenomena in Septic Tanks.

Type of Settling Phenomena	Description	Occurrence in Septic Tank
Type 1: Discrete Particle	Settling of particles in a suspension of low solids concentration, particles settle as individual entities, with little or no interaction with adjacent particles.	Removes heavier discrete irregular particles.
Type 2: Flocculant	Individual particles tend to coalesce, or flocculate, increasing their mass and settling rate.	Removes lighter particles that flocculate into heavier particles.
Type 3: Hindered (Zone)	The particles tend to remain in fixed positions with respect to each other, a solids-liquid interface develops at the top of the settling mass, which settles as a unit.	Occurs if biological floc develops.
Type 4: Compression	Consolidation and compression of sediment takes place from the weight of the particles which are constantly being added. Further settling can occur only by compression of the structure.	Occurs in the lower sludge mass.

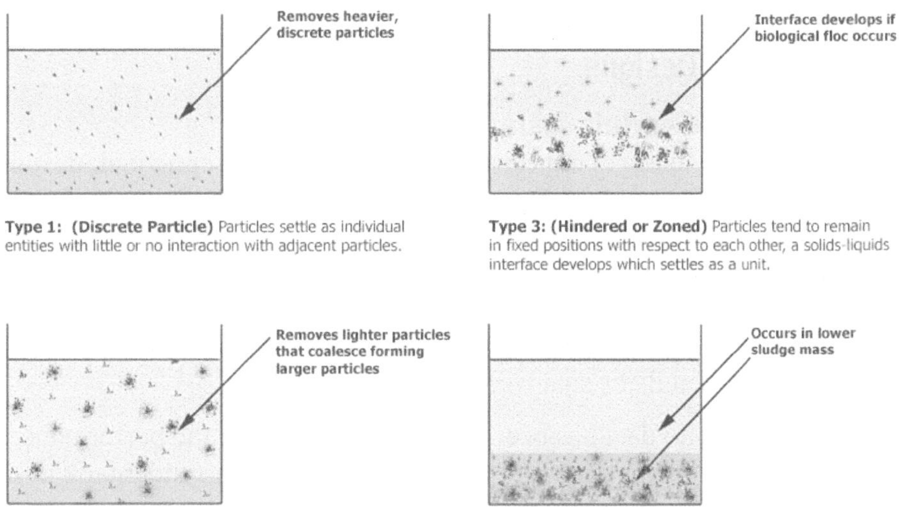

Type 1: (Discrete Particle) Particles settle as individual entities with little or no interaction with adjacent particles.

Type 3: (Hindered or Zoned) Particles tend to remain in fixed positions with respect to each other, a solids-liquids interface develops which settles as a unit.

Type 2: (Flocculant) Individual particles tend to flocculate, increasing their mass and settling rate.

Type 4: (Compression) Consolidation and compression of sediment take place from the weight of particles which are constantly being added.

Figure 3. Types of Settling Phenomena per Ideal Settling Theory.

Type 1 and 2 settling characteristics—the predominant settling phenomena in septic tanks—are functions of surface overflow rate, rather than depth per se. That is, a greater settling efficiency (more particles settled) can be achieved by maximizing the surface area (hence, reducing the liquid depth) for a given tank volume. In reality, because the contents in a properly-operating septic tank stratify into relatively distinct zones and because a design objective is to draw effluent from the clear zone, there is a trade-off between shallow depths (for greater surface area) and deeper depths (to maximize vertical extent of the clear, effluent zone).

Hazen (1904) recognized the hydraulic benefits of baffling in settling tanks to prevent circulation between compartments and maintained that the compartments will behave as multiple tanks, each with their own settling efficiencies. It follows that, per ideal settling theory, solids removal efficiency should be a function of the surface area of each individual settling compartment. In theory, for a given total volume (and surface area), single compartment tanks should outperform multiple compartment tanks. However, Hazen did not consider units within which both settling and digestion functions co-occur. In most cases where biologically mature units have been studied, two-compartment septic tanks have been shown to have better solids removal efficiencies than comparably sized single-compartment tanks (Laak, 1980; Rock and Boyer, 1995; Rich, 2006). This is generally attributed to the physical separation of the main settling and digestion compartment from the smaller outlet chamber where digestion is limited, thus isolating the impacts of biological digestion. In the first compartment or in single-compartment tanks, settling function may be impaired by rising gasses resulting from digestion in the settled sludge layer which resuspends particles.

While septic tank performance may deviate from strictly ideal settling conditions due to several other factors including inlet and outlet turbulence, eddy currents, thermal effects, solids resuspension and scouring, short-circuiting and fluctuations in influent flow rates, the presence of a significant scum layer and a lack of continuous sludge removal, its proper application can greatly inform engineering evaluations of alternate septic tank designs.

Alternate Septic Tank Designs

Meander and Closed-Conduit, Laminar Flow Tanks

J.H.T. (Tim) Winneberger conceptualized an enhanced septic tank design while a researcher at the University of California—Berkeley Sanitary Engineering Research Lab in the 1950s and 1960s. In Winneberger's design, which was later coined the "meander tank" by Warshall (1979), the septic tank would have longitudinally-placed baffle walls rather than baffles placed perpendicular to the flow, as is typical practice. Winneberger (1984) qualitatively described the operation of a pilot-scale 140-gallon rectangular tank, suggesting that it performed quite well, although no details or data were presented.

While essentially no data has been published comparing the performance of meander tanks versus other designs, many practitioners believe that the meander tank design has great merit and there are indications that such tanks are being used, but primarily for large, engineered system designs. Seabloom et al. (2005) state that the parallel chamber configuration is a way to increase length-to-width ratio, provide a more uniform hydraulic profile across the cross-sectional flow area, reduce short-circuiting, reduce inlet and outlet turbulent zones and improve overall tank effectiveness. Additionally, the serpentine flow pattern causes changes in the direction of flow which may enhance settling in low velocity zones adjacent to bends, similar to a meandering natural stream. On the other hand, a narrower cross-sectional flow area will yield greater velocities for a given surface area and depth potentially scouring settled solids and impacting solids removal efficacy. A schematic of a meander tank is provided as Figure 4.

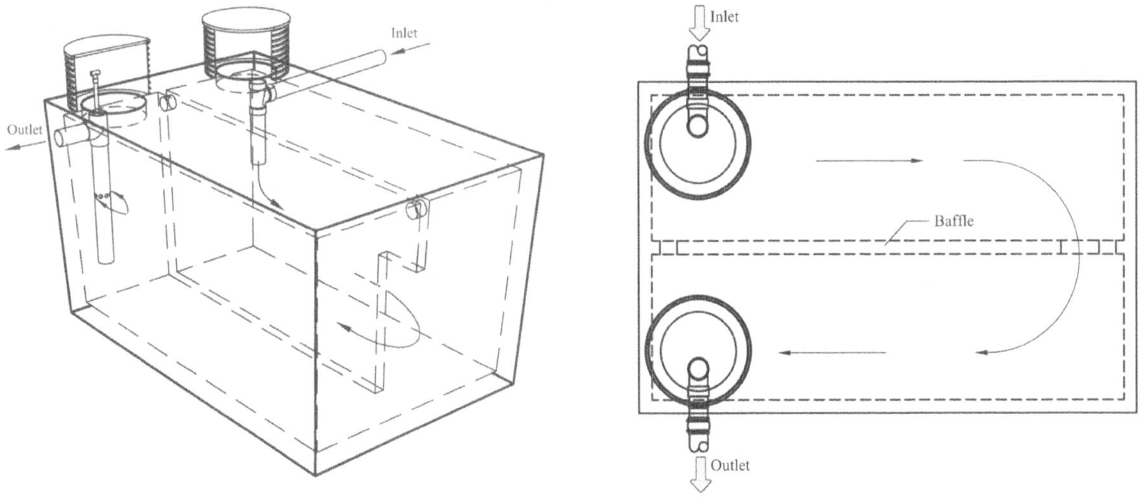

Figure 4. Isometric Schematic of Two-Compartment Meander Tank with Internal Pumping Chamber (courtesy of T.R. Bounds).

Jowett and Lay (2006) have extended some of Winneberger's ideas to develop prototypes of alternate tank designs. Like Winneberger, they question the need for airspace above the water level[1], suggesting that tanks would perform better with more sludge and less scum. They speculate that the scum layer is formed primarily from the biogas driven resuspension of settled solids and that the airspace itself contributes to the formation of a thick scum layer by allowing molds to grow on the surface (as described by Dunbar, 1908). They also suggest that effluent quality could be improved by encouraging laminar fluid flow by using long narrow tanks with little hydraulic dead space and relatively short inlet and outlet zones, similar to Winneberger's meander tank. They describe proprietary tank designs (WaterTube^{TM}) designed with no airspace (except in access risers) and a long, narrow, shallow aspect to induce closed conduit, laminar flow[2] (Figure 5).

According to Jowett and Lay (2006), a prototype tank was tested versus a conventional single-compartment tank design at the same loading rates, as confirmed by the Buzzard's Bay testing facility in Massachusetts. The conventional tank was reported to have accumulated more sludge and scum than the proprietary tank, yet had poorer effluent quality in parameters measured [total suspended solids (TSS), carbonaceous biochemical oxygen demand (cBOD), chemical oxygen demand (COD), fecal coliform (FC) and FOG] when tested at the same loading rates. The proprietary tank reportedly had scum only in the inlet riser and very little sludge halfway along the flow pathway (Jowett and Lay, 2006). A more detailed report of these tests is in publication and should be referenced when available for more information.

---

[1] Arguments against eliminating the air space include the function of the air space for providing surge capacity for smoothing out peak flows, for storage during power outages or failure of downstream dosing systems, and natural venting of gas byproducts.

[2] From a fluid dynamics (i.e., velocity/Reynolds number) perspective, tanks with a larger cross-sectional flow area would yield lower flow velocities and thus potentially more laminar conditions, given equivalent volumes. Long narrow channels, under the same hydraulic conditions, will force higher cross-sectional velocities.

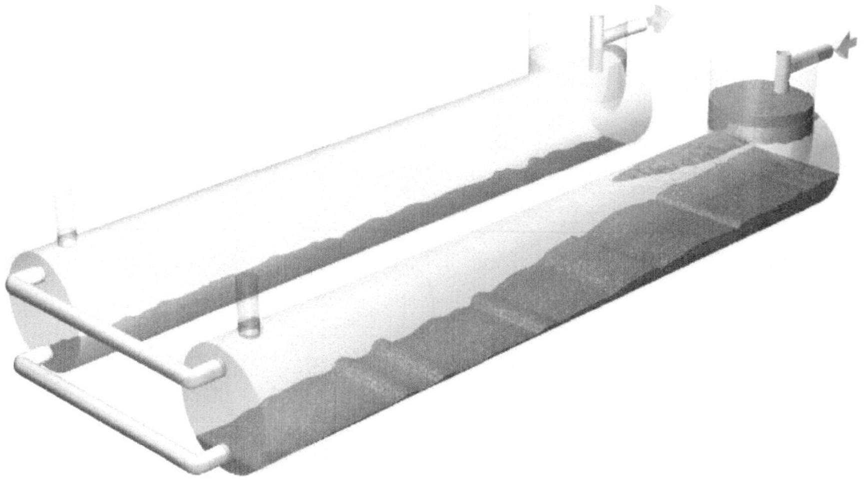

Figure 5. Isometric Schematic of Two-Compartment Closed Conduit Laminar Flow Tank (courtesy of Waterloo Biofilter).

UASB-Septic Tanks and Baffled Anaerobic Reactors

International researchers have led the development of technologies to enhance the anaerobic digestion capabilities of septic tanks. Centralized treatment in impoverished or developing tropical regions has often utilized upflow anaerobic sludge blanket (UASB) reactor technology, the introduction of which is generally attributed to Lettinga et al. (1980). On the decentralized level, hybrid septic tank-UASB reactors, sometimes referred to as anaerobic baffled reactors or baffled anaerobic septic tanks (BAST), are being developed and used (Nguyen et al., 2007; Koottatep et al., 2004). These technologies have been most popular in tropical and semi-tropical areas because their temperatures are more favorable for enhanced digestion and because many of these areas are economically depressed, with relatively little resources for advanced treatment systems (e.g., energy for aeration). However, research in temperate and cold climates has also been promising (Elmitwalli et al., 2003; Luostarinen and Rintala, 2006).

Conceptually, UASB-septic tanks have a series of compartments separated by baffle walls where effluent from one compartment is directed downward to flow up through the settled sludge blanket of the next compartment in series. A schematic of a four-chamber UASB-septic tank is provided as Figure 6. Several advantages are conferred by having the wastewater flow through the sludge blanket. Most notably, the blanket serves to retain solids better via enhanced flocculation, and passage of dissolved organics and nutrients through the biologically active sludge enhances their biological transformation and removal from the liquid phase. Relatively high ambient temperatures generally preclude the need to add heat, and digestion of accumulated solids is sufficient to minimize the need for solids removal. In less warm climates, biogases (principally methane) can be collected and used to generate supplemental heat for the reactor.

Nguyen et al. (2007) reported on laboratory studies comparing the performance of conventional two-compartment septic tanks and BAST units with and without internal anaerobic filters (AF) that can be used as an added barrier to solids carryover. COD and TSS removal efficiencies were significantly greater and standard deviations lower (implying better consistency) for the BAST as compared with comparably sized conventional septic tanks. The effect of hydraulic retention time (HRT) and number of stages was also evaluated and field retrofits using BAST and BAST with anaerobic filter (AF) units were studied.

The use of UASB-septic tank reactors may have merit, particularly in warmer climates of the U.S. where UASB-septic tank reactors could be used to enhance solids retention. It is important to note that UASB-septic tanks are gaining popularity in developing countries, implying that their capital and operation costs are not restrictive.

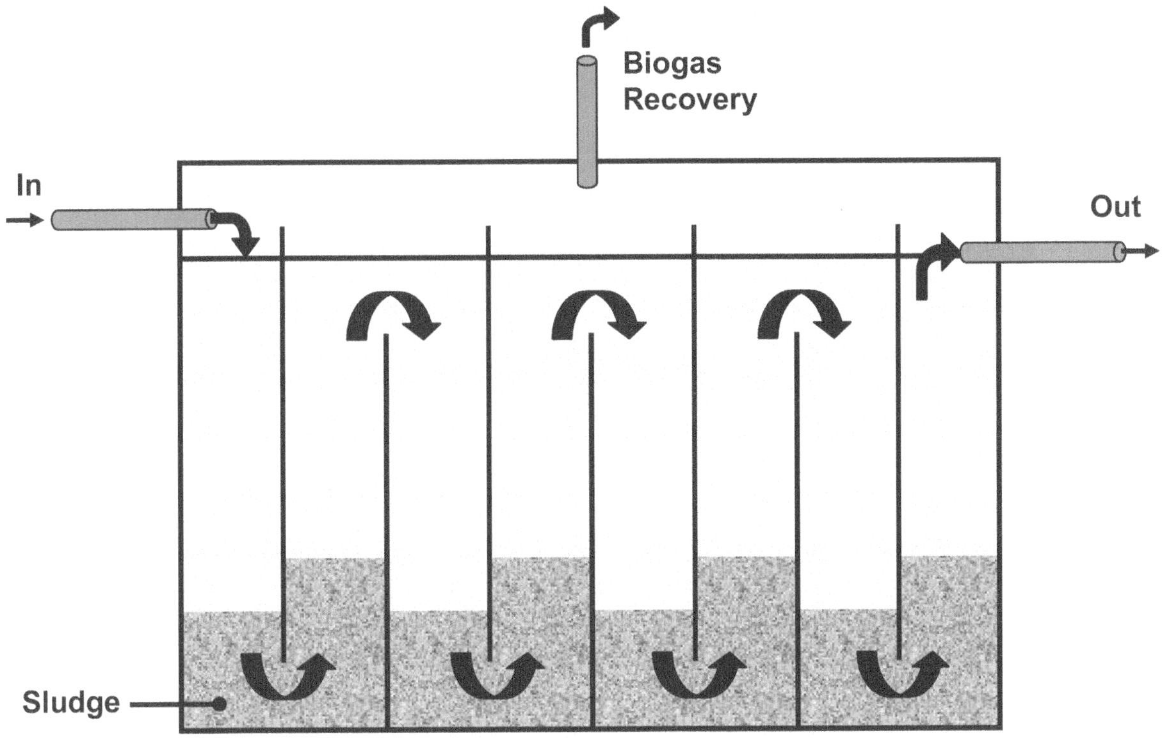

Figure 6. Profile Schematic of a Four-Stage UASB-Septic Tank.

Imhoff Tanks

The Imhoff tank was designed in 1906 by Karl Imhoff in Germany (Lens et al., 2001). The design improved upon standard single-compartment settling and septic tanks by separating the settling and digestion chambers. Imhoff tanks comprised almost half of the primary treatment systems in the U.S. by the end of the 1930s (Wolfe, 1999).

Imhoff tanks can be thought of as a two-compartment septic tank with compartments separated vertically. Unlike conventional two-compartment septic tanks, however, flow is only through one compartment—the top, or settling compartment. Settled solids slide down inclined walls into a lower digestion compartment (typically with a sloped hopper-type bottom) and internal baffling prevents biogas-driven resuspension of solids into the clear settling zone.

Imhoff tanks are by no means a new advancement of septic tank technology. Rather, these primary treatment units were widely-used in the early-to-mid 20th century, but have lost popularity in the U.S. (indications are that they are still relatively widely used internationally, most likely in systems larger than most decentralized systems in the U.S.) in the past 50 years. Reasons for the loss in popularity are unclear, but can be presumed to be due in part to the increase in popularity of the standard septic tank because of its lower cost and relative construction simplicity and subsequent codification of standard septic tank designs into state and local regulations for decentralized wastewater treatment.

Performance data for Imhoff tanks in decentralized systems is scarce; however, Baumann and Babbitt (1953) studied two Imhoff tanks alongside vertical cylindrical and rectangular multi-compartment septic tanks. The Imhoff tanks performed best for the removal of solids (suspended and settleable) under controlled pilot conditions. They also did not unload solids seasonally, as did the conventional tanks without effluent gas baffles. BOD removal, however, was lower in the Imhoff tanks, which the authors attributed to the prevention of clear zone seeding by gas bubbles evolved in the sludge layer of the Imhoff units.

Septic Tank Treatment Efficacy

Objectives

Primary treatment objectives of septic tanks include the reduction of solids (primarily the settleable fraction) and FOG from influent wastewater. While not necessarily a specific design consideration, organic carbon is typically removed to a significant extent via septic tank treatment. In addition to considering average concentrations of parameters of interest in STE, production of a consistent effluent is important; STE characteristics are almost always less variable than those for the associated raw influent.

The removal of BOD in septic tanks is predominantly through the removal of organic carbon associated with solids that exert an oxygen demand. There is some evidence that the clear zone of the septic tank acts as a biologically active zone, comparable to an activated sludge reactor, with the microbial activation occurring as a result of gas ebullition from the solids layer (Baumann and Babbitt, 1953). So, some removal of soluble BOD from the bulk liquid is undoubtedly occurring but the extent of this removal mechanism versus the removal of BOD through solids settling has not been established. Furthermore, it is clear that liquefaction of settled solids (i.e., stored BOD) liberates soluble BOD, thus competing with mechanisms which remove BOD from the influent. The contribution of these various processes does not appear to have been quantified.

STE quality standards have been proposed as an alternative to the prescriptive standards that currently dominate practice. Hoover et al. (1998) suggested a primary treatment standard (TS-1) which includes screened (i.e., for septic tanks fitted with effluent screening devices) and unscreened effluent quality criteria. Jantrania and Gross (2006) offer a similar suggestion. Gunn (1998) described efforts to develop such a standard in Australia which ultimately failed and settled on performance requirements pertaining to sludge and scum storage and pumpout and sizing criteria.

Nutrients

Septic tanks are generally not designed to remove nitrogen and phosphorus, although modest reductions will be realized from the removal of solids via settling. In addition, biological uptake of nutrients will also occur. However, because some of the accumulated solids will subsequently be digested, a portion of the associated nutrients will be resolubilized, restricting the total reductions achievable. The contributions of simultaneous nitrification-denitrification and ammonia volatilization in septic tanks do not appear to have been quantified.

Several efforts have been attempted over the years to optimize septic tanks for nutrient removal; however, there remain virtually no practical options for doing so, with the exception of recirculating nitrified secondary effluent to the septic tank for enhanced nitrogen removal (i.e., using the septic tank as an anoxic reactor).

Phosphorus removal in septic tanks is primarily a physical process with precipitation reactions contributing. Between 20 and 30% of the total phosphorus (TP) in raw wastewater is separated out in the form of sludge in a septic tank (Lombardo, 2006). Most of this phosphorus is particulate (1-3 mg/L or 20% of TP in the raw wastewater), although soluble orthophosphate may also be removed through mineral precipitation reactions, particularly vivianite (a hydrated iron phosphate) precipitation (Zanini et al., 1998). The potential of chemical phosphorus removal by dosing alum to septic tanks has been shown viable from a treatment perspective (Brandes, 1977a), but ultimately has not proven to be practically viable, primarily due to concerns about septage management (Etnier et al., 2005).

Other Wastewater Constituents

Relatively little information exists regarding the occurrence, fate and impacts of other constituents in septic tanks.

Metals are clearly present in raw sewage from households due to their presence in many household products and foods as well as interior plumbing systems. Parts-per-billion concentrations of various metals have been measured in residential STE in several studies. These studies suggest that metals generally partition to the solid phase via sorption and the formation of insoluble complexes in the septic tank and accumulate in septage (Sauer and Tyler, 1992).

Several studies have been conducted to evaluate the occurrence and fate of various synthetic organics in septic tanks. In one study, DeWalle et al. (1985), observed essentially no removal of priority pollutant volatile organic compounds (VOCs) in the septic tank; however, hydrocarbons were removed to some extent, with better removal with lower molecular weight. In several other studies (Greer and Boyle, 1987; Sherman and Anderson , 1991), toluene has been found as a very common constituent of domestic STEs. Viraraghavan and Hashem (1986) reviewed the existing literature and field sampled raw influent and STE at a household and detected six priority pollutant VOCs in the STE; however, toluene was measured in the raw influent but not the STE, in contrast to the other studies mentioned.

So-called "contaminants of emerging concern" have received a lot of attention over the past decade because of their almost ubiquitous occurrence in the water environment. A number of pharmaceuticals, personal care products and human metabolites have been detected in STE from domestic sources (DeJong et al., 2004; Brooks, 2006), but work to assess their treatability in septic tanks is on-going and not yet conclusive. Consistent with the observations of several practitioners, Hinkle et al. (2005) suspected that the long-term use of prescription drugs in households may have negatively impacted the performance of septic tanks in the La Pine, Oregon Project.

While the removal of microbes is certainly not an objective of the biologically active septic tank, the fate of pathogens in particular is of interest as there is some uncertainty about the viability of pathogenic organisms outside of their hosts and their persistence in abandoned septic tanks, for example. Otis et al. (1993) reported that septic tanks have not been found to appreciably reduce the bacterial numbers present in raw wastewaters. Removal of viruses in septic tanks is attributed to sedimentation as they are rarely free and isolated in the environment, but rather tend to be in aggregate form or linked with organic matter and other suspended solids. The occurrence of pathogens in STE appears dependent upon infection and excretion rates of the users. Indicator organisms (e.g., E. coli, FC) appear to be persistent in septic tanks. That is, septic tank liquors continue to have high E. coli or FC counts for weeks or months after infections cease or tanks are taken out of service (Anderson et al., 1991).

Several researchers have addressed the impacts of household chemicals on septic tank performance. Most have concluded that household chemicals used at normal or recommended levels do not appreciably impact septic tank performance, but that these same chemicals may have severe detrimental effects when disposed of down the drain or otherwise used at higher than recommended levels (Gross, 1987).

The development of reliable methods for assessing household chemical safety for septic tanks is potentially of great importance. Vaishnav and McCabe (1996) developed a 96-hour test protocol to assess the potential effect of consumer products on anaerobic sludge respiration in septic tanks. Edwards and DeCarvalho (1998) conducted a literature review of techniques to evaluate the toxicity of cleaning products to residential wastewater treatment processes including septic tanks, providing a good starting point for developing standard testing procedures.

Other, more general references help establish the effects of various elements and compounds on the biological health of septic tanks. Table 3 provides summary data on inhibitory concentrations of common contaminants taken from a variety of sources.

Table 3. Inhibitory Concentrations of Contaminants (courtesy of Orenco Systems, Inc.).

			Inhibitory Threshold		
Contaminant	Symbol	MCL	Aerobic		Anaerobic
		EPA DW	Heterotrophic	Autotrophic	Heterotrophic
		mg/L	*mg/L*	*mg/L*	*mg/L*
Aluminum	Al	*0.2 SCL*	15 - 26		
Ammonia-Nitrogen	NH_3-N		480		1500
Arsenic	As	0.01	0.05		0.1 - 1.0
Boron	B		0.5 - 10		2.0
Cadmium	Cd^{++}	0.005	1.0	5.0 - 9.0	0.02 - 1.0
Calcium	Ca^{++}		2500		2500
Magnesium	Mg^{++}			50	1000
Potassium	K^{+}				2500
Sodium	Na^{+}				3500
Chromium	Cr^{+6}	0.1 (total)	10 (total) 1 (hexavalent)	0.25 - 1.0	1.5 - 50
Chloride	Cl^{-}	*250 SCL*		180	
Copper	Cu^{++}	1.3	0.1 - 1.0	0.005 - 0.5	0.5
Cyanide	CN	0.2	0.05 - 20	0.3 - 20	0.1 - 4
Formaldehyde					50-200
Iodine	I		10		
Iron	Fe^{-}	*0.3 SCL*	5 - 500		5
Lead	Pb^{++}		0.1 - 10	0.5 - 1.7	50 - 250
Manganese	Mn	*0.05 SCL*	10		
Mercury	Hg	0.002	0.1 - 5	2 - 12.5	1400
Molybdenum	Mo				
Nickel	Ni^{++}		1.0 - 5	0.25	2 - 200
Silver	Ag	*0.1 SCL*	0.03 - 5	0.25	
Sulfate	SO_4^{--}	*250 SCL*			500
Sulfide	S^{--}		50		50 - 100
Tin	Sn				9
Vanadium	V		20		
Zinc	Zn^{++}	*5 SCL*	0.3 - 1.0	0.01 - 1.0	1.0 - 10

Compiled from: Grady et al. (1998)
Tchobanoglous et al. (2003)
Crites and Tchobanoglous (1998)
EPA (1987)
Vaishnav and McCabe (1996)
Bitton (1999)

Wastewater Source and Characteristics

Facility Type

Almost everything that is known about septic tank performance comes from studies of septic tanks serving single residential sources (i.e., households). Septic tank performance has been studied in detail for few other facility types and in fact, influent wastewater quality has been characterized for relatively few facility types, at least as reported in the literature. Additionally, because of the relative difficulty in obtaining representative raw influent wastewater samples, real-world paired raw sewage influent/effluent data for septic tanks is quite rare (usually, only STE data is reported), so the impact of specific wastewater characteristics on septic tank performance is difficult to determine with confidence.

Several focused efforts to characterize raw wastewaters from different sources have been made over the years (Siegrist et al., 1976; Otis et al., 1974). Most recently, Lowe et al., (2007) conducted a review of the literature characterizing single source wastewaters for WERF Project No. 04-DEC-1 "Influent Constituent Characteristics of the Modern Waste Stream from Single Sources", in preparation for a second phase of the project, whereby field sampling will be conducted to update existing information on the makeup of single-family residential wastewater sources. The main products of this project, so far, are cumulative frequency diagrams (CFDs) for various raw wastewater and STE characteristics based on data from the literature review. A CFD is a graphical depiction of constituent concentrations in relation to cumulative frequencies of occurrence, expressed as a percent. An example CFD is provided as Figure 7.

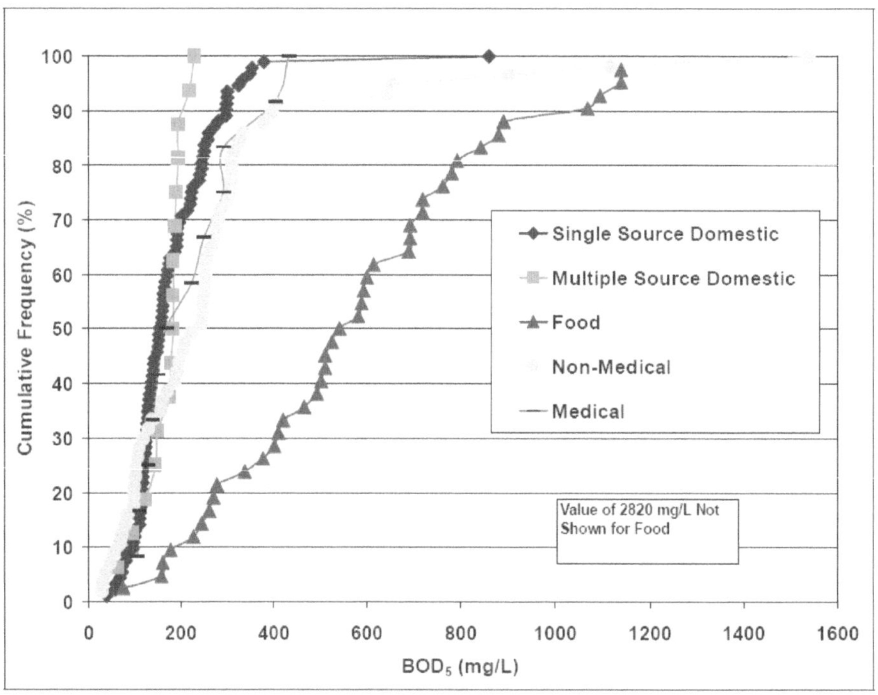

Figure 7. CFD STE 5-day Biochemical Oxygen Demand (BOD_5) by Source (Lowe et al., 2007).

While limited performance data for certain non-residential facility types does exist (STE data is relatively common, but paired influent/effluent data and primary treatment details are not), a cursory review of the data shows great inter-facility variability among like facilities, not to mention sometimes extraordinary intra-facility variability. Sampling frequencies for the studies that do exist do not appear to be sufficient to even broadly characterize primary unit performance with respect to facility type. Future research should include sufficient resources to allow sampling techniques and sample sizes that are likely to yield statistically meaningful results.

Oftentimes, primary treatment design enhancements for these non-residential and often high-strength facility wastewaters simply include adding tankage and effluent filtration. Such "enhancements" appear to be more based on convenience and perceived safety factors than on any real evaluation of how they may improve treatment for a given wastewater. An engineering study on scientifically-defendable primary treatment modifications (or not, if no practical enhancements exist) would help to validate design practice in the decentralized industry.

Besides single residential sources, a fair amount of septic tank performance information exists for residential cluster systems (further discussed in a subsequent section on Cluster Systems and Collection System Type) and food service facilities (further discussed in a subsequent section on Food Service Facilities and Grease Trap Design). Other facilities for which useful studies exist are listed in Table 4. Other facilities for which some information exists include campgrounds, dairy operations, manufacturing facilities, nursing homes, prisons and schools. The reader is referred to the bibliographic database for this project for more specific information.

Table 4. Summary of Relevant References on Septic Tanks Serving Non-Residential Facilities.

Facility Type	Summary	Applicable References
Vehicle Service Stations	Presents fate of heavy metals and VOCs in catch basins and septic tank serving publicly-owned service stations in Wisconsin	Sauer and Tyler, 1996
	Presents information on use and occurrence of solvents and other chemicals	Lueck and Shaw, 1994
Recreational Vehicle (RV) Dump Stations	Raw wastewater from dump stations had a high bacterial toxicity; also presents STE data from field samples	Matassa et al., 2005
	Studied effects of preservatives used in RVs on bench-scale septic tanks	Pearson et al., 1991
	Raw wastewater strength measurement	Pearson et al., 1980
	Raw wastewater strength measurement	Brown, et al. 1984
Highway Rest Areas	Paired influent/effluent data for septic tank	Griffin et al., 2002
	STE data for facilities with and without ultra low volume flush toilets	Otis et al., 1993
Funeral Homes (both studies commissioned by the National Funeral Directors Association)	Presents the results of a wastestream audit; septic tank impacts not specifically discussed	Killam Associates, 1995
	Study of the removal of formaldehyde and phenol by funeral home septic systems	LaKind and Bouwer, 2003

Cluster Systems and Collection System Type

Because of the more frequent use of alternative collection systems (STEP, grinder pump and others) in decentralized cluster systems, the impacts of collection system type on treatment efficiency and wastewater characteristics is of interest. Cooper and Rezek (1978) reported on the treatability of pressure sewage, focusing on the characterization and treatability of both grinder pump and septic tank effluent pump (STEP) system wastewaters. They report, logically, that STEP wastewater (which has been primary treated at each tank installation) is more dilute than conventional collection system wastewaters. Grinder pump wastewaters were reported to be 25-50% stronger and produce more finely divided solids when compared with conventional collection system wastewaters. Bounds (1982, 1988, 1993, 1996, and 1997b) has reported extensively on the pressure sewer system at Glide, Oregon. This has included an assessment of the wastewater characteristics as well as a study on sludge and scum accumulation rates. Bowne (1982b) reported that studies in Glide, Oregon, showed grinder influent BOD_5 was nearly three times greater than that of STEP influent and grinder TSS was over four times greater than that of STEP influent. Most of the STEP effluent at Glide during the study was not screened, as is currently widely practiced. Christopherson et al. (2006) compared the performance of septic tanks serving cluster systems having conventional gravity, STEP and grinder pump collection systems. While the data are limited, the trends are generally as expected, with STE from grinder pump systems being strongest in terms of solids and organics.

Source Separation

While source separation has been relatively uncommon in the U.S. (relative to some other nations), a fair amount of information exists regarding the performance of septic tanks for separated systems, where graywater - the wastewater from sinks, showers, laundry, and other appliances - is separated from that from toilets (blackwater). In the U.S., work was done in the 1970s to investigate the feasibility of graywater/blackwater separation (Siegrist, 1977a, 1977b, Siegrist and Boyle, 1981). In general, with the exception of nitrogen, STE concentrations were quite similar between graywater and blackwater septic tanks, although the graywater tanks had more dilute influents, so were not as effective in terms of percent reduction. Since blackwater contains the vast majority of the nitrogen in combined wastewaters, blackwater STEs were much higher in nitrogen than graywater septic tanks. Brandes (1978) found similar results, except that effluent phosphorus concentrations were much lower from the graywater septic tanks. The difference is most likely due to the use of phosphate-free detergents in the Brandes study. It is also notable that effluents from the graywater septic tank in this study contained coliform concentrations comparable to those from typical septic tanks treating combined wastewater.

Work continues to be done internationally where source separation is much more widespread; however, the information specific to septic tank performance is limited. Hellstrom and Jonsson (2003) investigated small treatment system technologies for use in Sweden, including urine and blackwater separation systems. While the focus of the evaluation was on advanced treatment technologies, performance results for septic tanks treating separated urine and graywater was summarized, suggesting comparable performance between systems, with urine separation septic tanks appearing to be more effective with respect to suspended solids (SS) and 7-day Biochemical Oxygen Demand (BOD_7) removals.

Ecological Sanitation (EcoSan) or Decentralised Sanitation and Reuse (DESAR) describe an internationally-driven decentralized wastewater management paradigm focusing on source separation and treatment and recovery and reuse of liquid, solid, nutrient and gas fractions. Technology development efforts are ongoing with these principals in mind. The concept of primary treatment in this paradigm is somewhat different than for the conventional, combined wastewater treatment predominantly practiced in the U.S. For example, urine, separated at the generation point, is typically retained in holding tanks for natural disinfection, which constitutes the treatment process. Likewise, feces are often dry composted, again precluding the need for conventional primary treatment. In place of septic tanks, the DESAR concept promotes the use of so-called "biogas latrines" and "communal biogas plants", where wastewaters are mixed with household waste to increase the potential energy output via anaerobic digestion (Lens et al., 2001).

Septic Tank Design Characteristics

Sizing

Sizing of septic tanks is addressed in most relevant texts and guidance manuals along with a variety of research papers, not to mention state regulations and industry standards. Early references and standards appeared to be based primary on easy-to-communicate, rule-of-thumb criteria that were passed along from source to source. Generally, these criteria were expressed in terms of nominal HRT (i.e., V/Q), with ranges from 6-72 hours cited. As the conceptual performance model (that is, development of distinct zones and storage of sludge/scum as an objective) for septic tank operation came to be developed, practitioners recognized that the effective HRT and nominal HRT would be different. For one thing, no reactors behave ideally; there is always some short-circuiting and unused volume. Secondly - and at least as important - it was recognized that the volume of a septic tank would become progressively more occupied with accumulated solids and, as such, the effective volume for settling, would decrease. Subsequent rules-of-thumb suggested an HRT of at least 24 hours after allowing for maximum sludge and scum accumulation, assumed to be 50% of the tank volume. Baumann and Babbitt (1953) clearly showed improvements in both solids and organics removal with increasing HRT in controlled pilot studies.

Starting with the PHS studies of the 1940s and continuing through more recent work by Bounds (1997a), sizing of septic tanks has become better defined with respect to the distinct vertical layers that will develop in a properly operating septic tank. It has been recognized that sizing is at least as important with regards to pumping frequency as it is in terms of settling efficiency. That is, a 24-hour HRT is more than enough to settle solids effectively (primary and secondary clarifiers in centralized systems may have HRTs in the 2-6 hour range); however, the capacity of the tank to store solids will be greatly limited with smaller sizes. Thus, the main advantage to a larger tank (i.e., longer HRT) is more efficient digestions, increased storage, and a longer interval between solids removal events, rather than improved solids removal per se (a potential consequence would be delayed maturation of septic tank biological function). The impact of tank sizing on pumping frequency may be dramatic and the selection of tank size can be considered a trade-off between higher capital costs for a larger tank and lower operation and maintenance (O&M) costs for less-frequent tank pumping (and vice-versa). Using sludge accumulation rate information along with equations for apportioning the various vertical layers in a properly operating septic tank, the designer can determine the size of tank required for desired pumping intervals (Bounds, 1997a).

For most facilities besides single residential, tank sizing is based on design flow for the facility. As such, the means of estimating or establishing the design flow is critically important with respect to septic tank sizing. Estimated flow rates for a number of different types of facilities are provided in various texts, guidance manuals, and state regulations, although it is generally accepted that flow monitoring from the facility in question (if existing) or a comparable facility is a more robust method for determining design flow rates, particularly for unique facility types. Two fairly recent studies sponsored by the American Water Works Association Research Foundation (AwwaRF) address residential and commercial end uses of water and are of particular note (Mayer et al., 1999; Dziegielewski et al., 2000).

Bounds (1994) specifically addressed septic tank sizing for large flows. As for residential systems, tank sizing calculations are presented based on apportioning volumes (which correspond to vertical depths) for various layers or zones. Based on sizing equations developed, for a pumping interval of two years, a tank capacity of approximately two times the average daily flow rate (2Q) should be provided. It should be noted that these calculations utilize sludge/scum accumulation rates established on a per capita basis based on residential STEP tanks. Therefore the criteria is more applicable to cluster (or other residential facility) system tank sizing than for other types of facilities, for which comparable sludge/scum accumulation rates have not been established.

Septic tanks are increasingly being used as anoxic reactors in biological nitrogen removal systems, where nitrified secondary effluent is returned back to the septic tank to facilitate denitrification using the organic carbon and low dissolved oxygen levels in the tank. In these cases, septic tanks minimally need to be sized to accommodate the resulting higher flows. Additional design enhancements may also be warranted, particularly if the tank is still to be expected to serve as a settling chamber and sludge digester.

Hydraulic Design and Characterization

Instantaneous flows are rarely considered in the design or sizing of tanks, but it is well known that surge flows may affect performance. Jones (1974) reported on hydraulic loading of and discharge from septic tanks serving individual residences. He reported that demand periods for domestic water use are typically less than five minutes in duration and established water demand frequency distributions for individual homes which were typically characterized by diurnal, or bimodal, peaking. The consequence for septic tank performance is that peak water demand tracks closely with influent flow. As such, septic tank design factors including inlet flow dispersion, surface area, surge storage, discharge rate and exit velocity can be considered in relation to instantaneous flows, with measures taken to optimize tank hydraulic performance and sludge retention. Features to improve hydraulic (and thus treatment) performance of septic tanks include compartmentation and increased surface area, among others.

The use of internal pump chambers in septic tanks presents a potential hydraulic interference that must be factored into design. Yost and Lingireddy (1997) measured velocities in the vicinity of the pump chambers and calculated Reynolds number in an attempt to compare local velocities with critical velocities for scour. They concluded that suitable pump vaults had an impact only locally around the vault inlets within 2.5 cm of the vault.

Computational fluid dynamics (CFD) is a branch of fluid mechanics that uses numerical methods and algorithms to analyze problems that involve fluid flows. CFD has been utilized within the wastewater field for simulating the performance of clarifiers, contact chambers and other processes where hydraulic conditions are particularly important and several projects have utilized or are using CFD specifically to analyze the hydraulic performance of septic tanks and grease traps. Additional CFD modeling and calibration considering the various functions of septic tanks and grease traps could lead to the development of tools for assessing and designing primary treatment enhancements and improved performance predictions, particularly when applied in concert with flow dispersion and related hydraulic studies.

Richardson et al. (2002) reported on the development of a simulation product that was used to predict the fate of flushable consumer products in septic tanks, for a project sponsored by Proctor and Gamble. Objectives were to understand flow patterns that develop within normally operating septic tanks, to determine settling properties of materials likely to enter septic tanks, and to identify additional data required to improve the understanding of the behavior of solids transport within septic tanks. The CFD model developed was validated via a field study conducted at the National Sanitation Foundation's (now NSF International) test site in Chelsea,

Michigan. The numerical model used to simulate flow through septic tanks was developed using the FLOW-3D software system. The model involved the solution of three-dimensional equations of motion (Navier-Stokes equations) and the use of a drift-flux algorithm to account for solids settling and density variations within the carrying fluid. The result of the effort suggested that simple settling tests could be used to predict the fate of products in septic tanks using calibrated CFD models.

McCorquodale et al. (2006) reported on the application of a CFD model to improve the performance of rectangular secondary clarifiers, using a two-dimensional model to assess performance versus various internal modifications. The CFD model was developed for a secondary clarifier for which performance data were available for calibration and incorporated 1) discrete, zone and compression settling, 2) flocculation, 3) non-Newtonian flow, 4) floatable particles, and 5) variable internal tank options including skirts and perforated baffles. While not specific to septic tanks or decentralized wastewater systems, this reference provides a good example of the types of performance-influencing factors that could be investigated using CFD modeling.

Ducoste (Principal Investigator for WERF Project No. 03-CTS-16T "Fats, Roots, Oils, and Grease (F.R.O.G) in Centralized and Decentralized Systems") has developed a preliminary CFD model to simulate a pilot grease interceptor and for use as a tool in developing alternative grease interceptor designs for the project.

Geometry

There is conflicting information regarding the effect of shape on septic tank performance. Early US PHS studies indicated that shape was not important given similar compartmentation details and capacities (Weibel et al., 1955; Troyan et al., 1984). Subsequent publications suggest, however, that rectangular (in plan view) tanks perform better than vertical cylindrical tanks (Rock and Boyer, 1995). Vertical cylindrical tanks are not as widely popular today as they may have been in the past. While not ideal from a treatment perspective, vertical cylindrical tanks do have advantages with respect to construction convenience and cost. By stacking cylindrical segments together making them deeper (also not ideal from a treatment or installation and watertightness perspective), a common diameter could be used to satisfy a variety of volumetric capacities. With respect to optimizing materials and structural considerations, cubic or spherical shapes would be most efficient for a given volume, although again not necessarily ideal with respect to performance.

Like those studies addressing tank shape, studies have not been able to clearly separate interior dimensions of septic tanks from other design variables that might influence performance. As indicated previously, all else being equal, septic tanks with larger surface areas should have greater settling efficiencies. Ingersoll et al. (1955) in an investigation of fundamental theories of sedimentation recommended that settling tanks be designed on the basis of surface area and that rectangular settling tanks should be long and narrow. Most texts, regulations and guidance documents recommend that long, narrow tanks (e.g., length-to-width ratios of 3:1 or greater) be provided in an effort to minimize short circuiting and increase the travel path of settling particles. Joy et al. (2004a, 2004b) describe a low-profile, horizontal, single-compartment septic tank developed in France, suggesting that the large surface area improves digestion by encouraging the diffusion of VFA into the clear zone, which keeps the pH of the sludge layer from becoming inhibitory. As such, they report significantly longer intervals between pumping as compared with conventional septic tanks. Jones (1974) also promoted the use of surface area as a key design parameter, with increased surface areas resulting in lower discharge rates and reducing the rate at which the depth of sludge and scum increases, minimizing rapid digestion upsets and associated solids carry-over. Ludwig (1949, 1950) demonstrated excellent performance with a battery of tanks in series forming a two-compartment tank with a 14.5:1 length-to-width ratio as compared with a standard two-compartment septic tank with a 2.5:1 ratio, at the same proportional loading rates.

Compartmentation

The question of compartmentation continues to generate disagreements among practitioners and academics. Nevertheless, the preponderance of the research—both controlled-pilot tests and field surveys—strongly suggests that multi-compartment tanks perform better than comparably sized single-compartment tanks under real-world conditions. Laak (1980) provided a comprehensive review of historical compartmentation studies and practices and concluded that multicompartment tanks performed better than single-compartment tanks for both suspended solids and organics removal. Sludge and scum, however, will accumulate more quickly in the primary compartment. Thus, while compartmentation may enhance solids removal, maintaining capacity in the primary compartment for solids accumulation and retention time for sludge digestion must be considered to maintain long-term performance efficiencies. The primary advantage of providing multiple compartments is to limit the majority of the digestion and settling disruption to a first compartment. Solids resuspended in the first compartment have additional, less-disrupted settling time in subsequent compartments. Other benefits of compartmentation include potentially improved hydraulic performance in terms of minimizing short-circuiting and flow modulation (i.e., the flow restriction caused by intercompartment transfer slows down the flow into the second compartment and outlet during peak flow periods) (Jones, 1974), although hydraulic performance could also potentially be enhanced via improved inlet and outlet hydraulic design (flow dispersion).

Most guidance calls for a first compartment that is as large as or larger than any subsequent compartments, as sedimentation efficacy is a function of surface loading rate (SLR) with lower SLRs yielding greater solids removal efficiencies. In tanks with compartments that are hydraulically connected, providing unequally sized compartments avoids oscillations between compartments that can interfere with settling. When surface areas of compartments are unequal, there is a dampening effect on movement of water between compartments because momentum and head are unequal (Baumann et al., 1977).

A variety of intercompartment connections have been used and tested for multi-compartment single tanks. While it appears that some sort of flow restriction device at the connection may be beneficial from a hydraulic standpoint, results of tests on intercompartment connection devices have been conflicting and, as such, inconclusive. Practitioners often suggest that a larger intercompartment connection is preferable, since velocities near and through the connector will be lower, reducing the possibility of scour and settling disruption. Rock and Boyer (1995) reported that a two-compartment tank utilizing a slotted transfer opening performed significantly better than a similar tank which used a 4" diameter hole. The difference was statistically significant for both BOD_5 and TSS, although it should be noted that the tanks were not operated in parallel, so these results could be confounded by other factors. Weibel et al. (1955) studied compartmentation details and found that a tank equipped with a 4-inch slot, widthwise across the baffle wall submerged 12 inches resulted in HRTs equivalent to an equal size single-compartment tank, but less than half that of an inverted U bend equipped three-compartment tank. Likewise the coefficient of dispersion of this inverted U bend tank was best followed by the slotted connection followed by the single-compartment tank. When a four-inch hole was used instead of a slot, the experimental detention time increased and a slight improvement in the coefficient of dispersion was also noted. Plate settlers have been used as intercompartment transfer devices, but what results are available indicated that effluent quality was comparable to that from tanks with conventional transfer devices (Simmons et al., 1981). Jones (1978) emphasized measures to prevent oscillatory mixing and intercompartmental mixing in multi-compartment tanks.

Where tanks-in-series are used to provide compartmentation, the tanks are not typically hydraulically interconnected; that is, the effluent from the first tank enters the second tank above its water line. The tanks thus behave somewhat independently. As such, oscillation between compartments is not a concern and interconnection details are usually as specified for the outlet of a single septic tank. Hydraulic interconnection of tanks in series is generally not recommended because tank connections beneath the operating water level of the tank are more subject to leakage than are connections above the water level.

Septic Tank Inlet, Outlet, and Appurtenances

As for many factors that likely influence septic tank performance, considering the great natural variability in septic tank performance (and, likewise, the difficulty in capturing that variability), it has been difficult to isolate the specific effects of inlet and outlet design on performance. Many design recommendations, therefore, are based more on theory than on verified performance. Modeling and simple flow dispersion testing could be better employed to guide further pilot- and then field-scale research in this area.

Jones (1974) suggested that increasing the size of the outlet flow area (e.g., effluent tee) will reduce velocity and decreasing the size of the outlet hydraulic control device (outlet pipe) will reduce discharge rate, both improving hydraulic conditions within a septic tank and best allowing for uninterrupted settling. Baumann et al. (1977) suggested that slopes of the sewer line to the septic tank should be restricted in order to minimize instantaneous flow rate into the septic tank. They also suggest that the inlet be designed to dissipate energy, preferably via a sanitary tee, an elbow, or a specially designed influent device. Jones (1978) provides specific recommendations on the inlet elbow radius and vertical extension depth for maximizing energy dissipation and minimizing turbulence, but no supporting data were provided for confirmation. The PHS (Weibel et al., 1955) reported that a tank equipped with a tee inlet had a lower coefficient of dispersion and as such, a better flow pattern than a tank equipped with a straight inlet. Whether any of these factors would have an appreciable influence on performance is unknown.

Despite their increasing popularity and importance to primary unit performance today, effluent screens have not been extensively independently-tested and reported in the literature. Undoubtedly, testing has been conducted by manufacturers and suppliers, but, for the most part, the results have not been reported widely in peer-reviewed forums. However, there are in existence hundreds of decentralized effluent sewer systems that require and use effluent screens on both gravity [septic tank effluent gravity (STEG)] and pumping (STEP) units. These municipal utilities monitor and report influent characteristics per permit requirements. While these results are not widely distributed, they are typically third-party and may be available through contact with the municipality or the regulatory authority.

Byers et al. (2001) presented work primarily focused on developing methodologies for testing effluent screens. Since a main objective of effluent filtration is the removal of larger, neutral buoyancy solids, a gross solids sampler device was designed and tested. Testing of screens using this apparatus with wastewater heavily loaded with paper and large particulates, showed screened and unscreened collected solids masses of 2.52 and 15.5 grams respectively (difference statistically significant at a p-value of 0.01).

Bounds (1994) addressed the issue of modulating peak flows through primary tanks using effluent screens whose discharge rate is controlled by their flow orifices. Bounds (1997a) presented performance data showing that STE concentrations of BOD_5 and TSS for tanks without effluent screens averaged 156 and 84 compared with averages of 133 and 30 mg/L for BOD_5 and TSS for screened effluents.

Like effluent screens, outlet gas deflection devices at one time enjoyed more popular use, being described in a number of relevant references; however, no comparative performance data for tanks fitted with and without these devices was uncovered.

Food Service Facilities and Grease Trap Design

Little is known about the factors impacting the performance of grease traps and likewise it appears that almost nothing has been done to optimize design or to confirm that sizing criteria in use today (which vary widely) are appropriate. Kommalapati and Johnson (2001) conducted a literature review on design parameters for modern grease traps and high strength wastes and reported mostly background information on greases and grease removal. The primary conclusions of their literature review were that there are virtually no reports available regarding the design of modern grease traps and that a comprehensive study is needed. The Florida Department of Health (1996) conducted a similar literature review during the first phase of a study of failure of decentralized systems serving food service establishments. Additionally, preliminary sampling (typically from the final dosing tank of the system) was conducted showing uniformly high levels of BOD_5, oil and grease (O&G) and TSS along with high variability. They report that the results appeared to be more closely related to the design of the septic tanks and grease traps than the type of restaurant. Plans for a Phase II study to investigate specific design factors influencing effluent strength were presented, but the results of this work do not appear to be available.

What is consistent in the grease trap literature is that large outdoor grease traps similar in design to septic tanks (with some logical modifications to enhance grease retention) are necessary to remove substantial amounts of FOG in decentralized wastewater treatment systems treating food service facilities. Additionally, it is clear that management and operating practices of restaurants are critically important grease control measures. Practitioners agree that blackwater should only be co-mingled downflow of the grease and oil separation units, so that removal of accumulated materials for rendering is viable.

In a study by Siegrist et al. (1984a, 1984b), concentrations of organics and solids in STE from food service facilities were found to be significantly greater than from domestic sources. For non-restaurant food service facilities, organic strength was found to be moderately higher than for domestic sources, but somewhat lower than domestic for other parameters. Although the performance of systems with and without grease traps was not explicitly compared, it appears that those with grease traps performed better, particularly with respect to O&G removal, than those without grease traps. Of the two restaurants with outdoor grease traps, STE O&G averaged 65 and 47 mg/L, while for the remaining four restaurants (two of which had indoor grease interceptors and two of which had no grease interceptor or trap), STE O&G averaged 101, 40, 141 and 101 mg/L. Effluent TSS results show similar relative trends, but BOD and nutrient concentrations in the STE are inconsistent.

The City of Austin, Texas conducted a study of factors affecting grease trap performance (Hawk, 1982) in what appears to be one of the only published studies of its kind. The study included field sampling at four restaurants (each representing a different type of restaurant) as well as a controlled study of flow dynamics through improved prototype interceptor designs. They concluded that insufficient minimum detention time, influent grease and oil overloading, insufficient cleaning and emulsification due to elevated temperatures, detergent usage and mechanical shearing all adversely affected grease trap performance. While inlet flow deflection baffles improved hydraulic performance, actual HRTs were far less than calculated based on flow and volume. Shallow, long, high surface area designs performed best.

Wong et al. (1998) measured reductions of 44-57% in TSS and O&G across effluent screens at four restaurants in Tennessee.

WERF Project No. 03-CTS-16T "Fats, Roots, Oils, and Grease (F.R.O.G) in Centralized and Decentralized Systems", is being conducted by a research team led by Dr. Joel Ducoste, based at North Carolina State University. Its objectives are to:

- Determine the optimal design, sizing, and operations and maintenance criteria for grease interceptors
- Evaluate the effects that combined waste and waste from garbage grinders have on grease interceptors
- Develop a companion report that can be submitted to International Association of Plumbing and Mechanical Officials (IAPMO) for possible inclusion into the Uniform Plumbing Code (UPC).

A series of related papers characterize restaurant wastewaters and relate wastewater strength to restaurant management practices. Lesikar et al. (2004) calculated design values for BOD_5, TSS, FOG and flow rate using a trimmed dataset and found the parameters to be substantially higher than those design values reported in texts and guidance documents. Lesikar et al. (2005) correlated this data with restaurant management practices and cuisine type. Statistical analyses indicated that only three variables were significant at a $p\text{-value} < 0.05$: Cuisine Type, Presence of a Self-Serve Salad Bar and Number of Seats.

Seasonal/Temperature Effects on Biology

Digestion in septic tanks (and other wastewater systems) has been positively correlated with temperature (Pearson 1986, Boyer, 1992). Pearson (1986) analyzed the PHS data from the 1940s-1950s studies with an eye toward seasonal performance and the benefits of compartmentation. According to his analysis, effluent TSS from a single-compartment septic tank was about twice as high in summer as in winter. Multiple compartments alleviated summertime deterioration of suspended solids, presumably by isolating the zone of highest digestion from the outlet.

The well-understood dependence of biological reaction rates on temperature has also been clearly observed in laboratory and field observations of operating septic tanks. Numerous researchers have anecdotally described a "spring turnover" or "boil" and often an increase in effluent solids concentrations during warmer months. To summarize, settling and solids removal efficacy are optimized during the cooler months coincident with a growth in the amount of accumulated solids. In the warmer months, digestion and solids reduction are optimized, reducing the amount of accumulated solids, but gas ebullition during the increased digestion hinders settling and limits solids removal efficiency. Popular enhancements to septic tanks including compartmentation and the use of effluent devices that deflect rising bubbles and screen out relatively large neutral buoyancy solids, may help alleviate this seasonal drop off in solids removal efficiency.

Regional variations in the performance of septic tanks can be significant. In particular, septic tanks in warm climates may exhibit vigorous digestion year-round. In Florida, septic tanks that had been in operation for as long as 12 years and had never been pumped had relatively small accumulations of sludge (Sherman and Anderson, 1991). These tanks appear to be very efficient digesters, but relatively poorly functioning settling units. Regional and/or seasonal design modifications could be considered. More difficult settling conditions but relatively small accumulations of solids in warmer conditions suggest that tank sizing (for storage of accumulated solids) is not as important as are features to improve settling efficacy and minimize the effects of gas ebullition and resuspension of solids. Conversely, in cooler climates, conservative tank sizing (for solids storage) may be relatively more important than enhancing solids retention via design modifications.

The biological community of operating septic tanks is complex. The biological treatment occurring in operating septic tanks has typically been attributed to bacterial and fungal populations, although recent studies show that protozoans are ubiquitous and play an important role (Nair et al., 1999, Grimes 2007). Other higher level organisms (e.g., oligochaetes, nematodes) may also be found and play a significant role in septic tank treatment. Recent studies documented that the bacterial populations in septic tanks were genetically different from the contributing fecal matter (Gordon, et al. 2002). Likewise, the genetic bacterial diversity in septic tanks was lower than that of the household. Only a quarter of the measured genetic variation was attributed to temporal effects and each tank had a unique biological bacterial profile. Studies report that following peaks in bacterial growth, a subsequent protozoan population peak follows (Nair et al., 1999, Grimes 2007). Once a protozoan peak occurs, as in other systems, the bacterial population plummets, following classic predator-prey population cycles. What might appear to be temperature-related or other effects, may actually be due to population shifts of these complex ecosystems. In laboratory experiments using culture flasks, Grimes (2007) found that cyst-forming ciliate protozoans in samples of sludge from septic tanks were slow to excyst when food availability conditions were favorable. The time lag associated with excysting could be a significant factor for tanks that are subjected to periods of dormancy (seasonal housing, for example). Further studies are necessary to understand the development of these ecosystems in relation to tank age, household practices, seasonality, and other factors.

The published literature contains several references addressing the impacts of extreme cold conditions on septic tank and system performance, and measures for mitigating potential impacts. Wallace and English (2003) presented a paper which established a thermal heat loss and insulation calculation method for buried infrastructure, particularly septic tanks. They recommend insulation for tanks installed in areas where frost depths are at or below the elevation of the top of the buried tank.

No seasonal performance information has been found for grease traps; however, cooler ambient conditions should result in somewhat more effective heat transfer (dissipation) and theoretically improved separation. Since desired grease trap function is purely a physical, rather than biological, process, the inhibition of biological activity during cooler periods would be an additional benefit. Seasonal operational strategies do not appear to have been documented in the literature.

Sludge/Scum Accumulation and Removal

Sludge and scum accumulation rates have been established by a number of researchers over the years, beginning with the US PHS studies (Weibel et al., 1955). A list of the known studies on sludge and scum accumulation is provided as Table 5. In general, the results generated by these researchers have been in fairly good agreement and the relationships established can be used as a general guide for designers and planners. As indicated previously, pump-out frequency will be a function of septic tank size and design characteristics and a variety of facility-specific considerations, including number of users (family size for single residential sources) and water use and solids generation characteristics. The consistent theme from these investigations is that the typically recommended pump-out interval of 3-5 years[3] is often quite conservative and, more importantly, to the possible detriment of the system since methanogenesis does not fully develop until after about three years of operation. Philip et al. (1993) summarize the biological development (or "maturation") that occurs in septic tanks and its effect on system performance. An understanding of this fundamental biological evolution of septic tanks is critical to understanding their function and, hence, being able to improve their design and operation. Additional work could be conducted to develop methods to better assess sludge biology in the field. Furthermore, pumping protocols could be considered in light of a more holistic understanding of septic tank biological function.

Table 5. Summary of Relevant Studies on Sludge and Scum Accumulation Rates.

Reference	Summary
Weibel et al., (1955)	Results of field sampling of 129 septic tanks across U.S. in 1950s
Schmidt (1976)	Reference not collected.
Troyan et al., (1984)	Results of field sampling in Perth Australia yieded 47 liters/user-year average
Winneberger (1984)	Results of two-year studies in twelve septic tanks in Novato, California
Mancl (1984)	Used results of previous studies to estimate pumping frequency based on tank size and number of persons in residence
Ollivant (1993)	Results of audit of 1,125-tank STEP system in Montesano, Washington in 1988
Philip et al., (1993)	Results of three-year field study on 33 residential systems in Montpellier, France yielded 0.2 liters/user-day accumulation rate, requiring 5 year or greater pumping
Bounds (1997b)	Results of eight-year audit of 450 septic tanks in a STEP system in Glide, OR in 1980s

[3] To provide some historical reasoning with respect to the 3-5 year pumpout recommendation, early versions of the EPA's Design Manual for Onsite Wastewater Treatment and Disposal Systems discuss operation and maintenance and suggest that tank inspections should occur at intervals of 2 years. It states that if inspections are not carried out, a pump out frequency of 3-5 years is reasonable. However, it also indicates that the frequency can be adjusted accordingly and that inspections were the only way to determine definitively when a given tank needs pumped. In practice, inspections were not typically performed and regulatory enforcement entities at the time were typically not staffed to ensure compliance; therefore a conservative 3-5 year interval was recommended to ensure a high confidence level.

Use of Additives

Septic tank additives have been marketed as long as there have been septic tanks (Minnis and Burks, 1994). Commercial additives increased in popularity after World War II and today, at least 1,200 products are on the market (Friedman, 2007). Relative to the large number of products on the market, very few controlled studies on specific septic tank additives have been reported in technical forums.

Long (1997) described the State of Washington's program for assessing the safety of septic tank additive products, regulated under authority of the Revised Code of Washington 70.118 and Washington State Administrative Code 246-273. The state reviews the products' ingredient lists for products proposed for use as septic tank additives in the state based solely on evaluation of possible harm to public health or water quality when the additives are used as directed. It does not test the benefits of the products nor their performance with respect to claims made by the manufacturer. Nearly every product evaluated by the state has met the specified safety criteria.

In an industry response to Washington State's proposed regulation, Scow (1994) performed a review of the literature, including published scientific data as well as information provided by manufacturers and distributors of additives, in an effort to evaluate the efficacy and potential for adverse impacts of biological additives in septic systems. The author reports that there is little published information specifically on bioaugmentation in septic tanks. Several in-house research studies with mixed results were reviewed and are described. Jantrania et al. (1994) reported on the evaluation of one company's additive for improving septic tank operation under stress conditions caused by adding household chemicals. The evaluation included a pilot-scale experiment fed with wastewater from a municipal wastewater treatment plant. Under stressed conditions, STE O&G concentrations were approximately 40% lower (significant at 95% confidence level) for the additive-amended septic tanks and TSS concentrations were lower and less sludge accumulation was measured in the additive-amended tanks when subjected to highly stressed conditions.

One of the most commonly cited field studies on biological septic tank additives is a master's thesis from North Carolina State University (Clark, 1998). Results showed sludge depth and scum thicknesses to be controlled by their initial pre-additive values; none of the treatment sludge depths were significantly different than those of the controls. BOD_5 showed some minor treatment significance while TSS concentrations did not. Again, pre-additive values of BOD_5 and TSS appeared to have more control over subsequent measurements. The additives also did not appear to significantly influence the organism population in the tanks over the course of the study. Although all three additives tested contained live, viable organisms throughout the study, their microbial counts appeared to decrease significantly by the end of the 13-month study period. More recent related work by Pradhan et al. (2006) indicates that biological additives had no effect on bacterial populations and sludge, scum and total solids levels in septic tanks.

Baking soda (sodium bicarbonate) and alum are two potential additives that have been studied and shown to have positive effects on septic tank performance. Baking soda acts as both a flocculent, aiding settling, and a buffer, preventing pH depression and inhibition of digestion in the sludge layer (Laak et al., 1975). Alum use for phosphorus precipitation, while effective, is not a preferred phosphorus control option because of septage management issues. Philip et al. (1987) promote the use of mineral adsorbents with high specific areas for septic tank biological starters which fix bacteria and enzymes and accelerate digestion.

The bulk of the information on additives suggests that bioaugmentation products are generally benign but of marginal benefit. Chemical additives should be avoided due to their potential for harming tank biology and contaminating the receiving environment.

Water Softeners

Water softeners are cation exchange systems which adsorb calcium (Ca) and magnesium (Mg) from the influent water, displacing sodium (Na) in the process. When the ion exchange material approaches saturation, it is regenerated by flushing with a sodium chloride brine solution. The regeneration process results in a liquid stream containing the displaced Ca and Mg salts, along with the excess Na, as chlorides (Tyler et al., 1978).

The U.S. PHS (Weibel et al., 1955) first reported on studies investigating the effect of zeolite softener salts on septic tanks. They loaded a laboratory septic tank with a mixture of calcium, magnesium and sodium chloride. No adverse effect on tank performance was noted, but the salt concentration increased significantly near the bottom of the tank during colder weather. With the advent of warmer conditions and increased digestion, the tank purged itself of the accumulated salt. An investigation into the viability of sludge taken from the tank receiving the salt demonstrated that the microbial population had become acclimated to the saline environment. Conversely, companion bottle experiments showed that "a 1.2% mixture of sodium, calcium, and magnesium chlorides had a pronounced inhibiting effect on digestion of sludge-sewage mixture".

The Water Quality Association (WQA), through their research arm, the Water Quality Research Council, sponsored two studies on the effects of water softeners on septic systems in the late 1970s. The WQA uses these studies, discussed below, to argue that water softeners pose no problems for septic tanks although some practitioners cite anecdotal evidence that suggests that the regenerant brines impact septic tank biology.

In an effort to determine the effects of water softener use on septic tank function for the WQA, Tyler et al. (1978) conducted a literature review and sampled STE from household septic tanks. The salt concentrations (m_o) and sodium absorption ratios (SAR) of effluents from septic tanks connected to houses with water softeners were generally higher than those serving houses without water softeners although concentrations varied greatly. The osmotic potentials of STE from households with water softeners averaged -0.51 bars while STE from households without water softeners averaged -0.36 bars which the authors suggest is not significant since other studies had found that bacteria divide and grow most rapidly at an osmotic potential of -14 bars, although the referenced salt tolerance studies used media that bore little resemblance to STE. The authors recommend additional studies to determine salt concentrations in various zones of septic tanks and to determine the effects of salt concentrations on microbial activity in media resembling STE as well as the effects of pulses of high and low salt loadings to septic tank microbes.

NSF International performed a study for WQA in 1978 to investigate the effects of home water softener waste regeneration brines on individual aerobic wastewater treatment plants. The study found no statistically significant differences between BOD_5 and TSS data for control and test treatment units; however, since the tests were conducted on aerobic treatment units, extending the results to anaerobic septic tanks is not possible.

Bounds (1997b) suggests that water softener backwash brines may elevate chloride concentrations to levels that are toxic to septic tank microbes, increasing the rate of solids accumulation and potentially causing solids to remain in suspension due to ionic polarization caused by the heavy metallic salts. In problematic tanks on water softeners, Bounds observed minimal scum development, greater solids and grease carryover, a less distinct clear zone. He further suggests that improperly operating water softener systems cause problems for decentralized wastewater treatment systems, but that the effect of a properly-operating water softener system is open for debate.

The 2006 National Onsite Wastewater Recycling Association (NOWRA) conference in Denver, Colorado included a panel forum addressing the impacts of water softeners in decentralized wastewater treatment systems. Representatives from the WQA and NOWRA spoke and introduced a residential field survey study that they hope will shed light on the extent and causes of problems that have been observed in the field.

Garbage Grinders

The effect of garbage grinders on septic tank performance had first been studied by the PHS in the 1950s. Weibel et al. (1955) reported that ground garbage increased the solids load to septic tanks considerably and that STE is likely to contain more total solids than when sewage only is treated. Compared with septic tanks serving homes without garbage grinders, septic tanks serving houses with garbage grinders had higher sludge and scum accumulation rates. Field studies conducted in Duval County, Florida, showed increases in average annual sludge and scum accumulations in septic tanks receiving ground garbage, with most of the increase in the scum accumulation. Septic tanks receiving ground garbage also had slightly depressed pHs during the study period. Bounds (1997b) also found that in-sink garbage grinders added significantly to the scum layer in septic tanks in the Glide, Oregon STEP system audit. Although not specific to septic tanks, Haseltine (1950) provided a good summary of the effects of garbage addition to sewage and treatment systems, including its effect on sedimentation. Consistent with the findings above, increased scum formation was cited and the use of effluent screening devices suggested. A number of state regulations and several industry standards require more conservative septic tank sizing where garbage grinders are used. An allowance of 50% over standard sizing is typical.

Inspection, Operation and Maintenance

Septic tanks are generally low maintenance units. Nevertheless, septic tanks and grease traps must be inspected on a regular basis. There are countless numbers of fact sheets and other documents addressing the inspection of decentralized system including in-service septic tanks.

Even basic inspection and maintenance of decentralized wastewater treatment systems has repeatedly been identified as perhaps the most significant shortcoming limiting their sustainable, long-term operation. As such, there has been an increasing focus on products and methods for enhancing inspection and monitoring of septic tanks and these are the focus of the discussion below. However, basic inspection, operation, and maintenance information is critically important and can be found in Loudon et al. (2005). A fundamental concern is that of providing adequate access to tank contents for inspection and maintenance. Accordingly, risers over the access openings in the tank should be brought to finished grade to facilitate proper oversight activities.

Performance Monitoring and Troubleshooting

Russell and Bigelow (2002) presented information on their ultrasonic tank monitoring devices, with several brief case study examples on how the SepticWatch unit was used to detect and diagnose system malfunctions and other conditions. Worldstone, Inc. offers several models of automatic monitoring devices for use in both grease traps and septic tanks.

Christopherson et al. (2006) compared the use of common devices to measure the accumulation of thick, heavy grinder pump septic tank sludge and found that sludge depth measurements varied according to the device used as well as the rate at which they were lowered into the septic tank. Despite standardization of methods, sludge depth measurements still varied and the researchers utilized the average of triplicate samples for their results. They also developed a scheme to characterize sludge for consistency, color and shade, as indicators of relative biological activity in the tank. They indicate that septic tanks with adequate anaerobic digestion typically have thin, black sludge in the final compartment and that if sludge at a given sampling point is light in color and thick in consistency, anaerobic digestion is probably not occurring there. Carroll (2005) suggested similar criteria in a study that describes system troubleshooting in Iowa. A homeowner questionnaire was developed and used as a troubleshooting aide.

Philip et al. (1993) suggested that sludge accumulation rate is an appropriate indicator of septic tank biological health and digestion efficiency, with lower accumulation rates best. In their study, high accumulation rate tanks had higher effluent COD and total solids (TS) than low accumulation rate tanks. Paradoxically, they also report lower ORP (more highly anaerobic conditions) in the lower performance tanks than in the better performing ones (-339 versus -310 mV).

Other researchers have developed methodology to assess the biological function of septic tanks. For example, microscopic investigations of influent, effluent, and sludge layers in tanks reported as "dead" were conducted with a newly designed field kit (Grimes, 2007). The evaluation found that there was usually a physical reason for the failure. Excess food, insect exoskeletons, fibers from paper towels, and other physical disruptions were found upon microscopic examination. Additionally, Bounds et al. (2004) provide a list of factors that can be used to qualitatively assess treatment performance.

Sampling

Septic tank sampling can be challenging due to a number of factors. The difficulty in collecting representative samples of real-world raw wastewater (e.g., septic tank influent) is well-known. Raw single source wastewaters are particularly difficult to sample and characterize effectively as flows and contaminant loadings may vary widely over short periods of time. Due to the wide variability in water use in facilities such as homes, it is obvious that grab sampling will not capture the variability in influent characteristics; some form of compositing must be used. Lowe et al. (2007) have recently developed a device and methodology for raw wastewater sampling that updates the techniques developed and used in earlier characterization studies (Siegrist et al., 1976).

STE is considerably easier than raw wastewater to sample, as it is generally free of large solids and because the effluent represents a composite of the tank's liquid contents. Despite this, it is often advised to stagger paired effluent and influent sampling to account for residence time. Christopherson et al. (2006) compared 1-, 2- and 4-hour composite samples of septic tank supernatants but did not discern any significant differences between their BOD and TSS results (potentially due in part to a limited dataset).

In a study that primarily evaluated the field performance of one- and two-compartment septic tanks and advanced nitrogen removal systems, Haldeman et al. (2004) provide excellent information about sampling decentralized treatment systems. The authors suggest that grab sampling does indeed provide nearly identical results for most STE quality parameters as does composite sampling. The research team used dedicated Teflon bailers for sampling septic tanks with gravity discharges and dippers for sampling tanks with internal pump vaults.

Venting

The National Fire Protection Association classifies vented residential septic tanks as non-hazardous areas with respect to the National Electric Code (NFPA 820: Standard for Fire Protection in Wastewater Treatment and Collection Facilities). Winneberger (1984) reported data that is convincing and conclusive with regard to hydrogen sulfide concentrations that are generated within septic tanks being too low to be combustible and the methane-to-oxygen ratio being too high to be explosive. The atmosphere above the scum mat, relative to tanks that are adequately vented back through the building sewer and house vent, was reported as being nearly identical to normal atmospheric conditions. In order to avoid gas accumulation in pockets, septic tanks should be vented either through the house plumbing system, the drainfield or through screened atmospheric vents from the tanks themselves. This information notwithstanding, the interior of septic tanks and grease traps are confined spaces and should always be treated as such, with proper precautions taken during servicing (Pettit and Linn, 1987).

Tank Construction and Installation

Despite the limited amount of published information regarding septic tank construction and installation practices, most field practitioners agree that tank construction and installation are critical and that improper construction and/or installation has been responsible for many system malfunctions. As such, an emphasis on this element of tank "performance" is warranted. Construction and installation practices for tanks of various materials installed under various conditions are generally known; that is, fundamental research is not likely to be warranted. However, the further development of procedures for ensuring proper construction and installation—QA/QC—is an area needing work.

In the interest of brevity, discussions of tank construction installation, watertightness and corrosion are not presented in detail here. The reader is referred to other excellent references (e.g., Loudon et al., 2005) for information on these important issues. In particular, watertightness is one of the most important attributes of properly functioning primary tanks. There is extensive anecdotal evidence that "leaky tanks" are a major problem for decentralized wastewater treatment systems. NOWRA's proposed model code contains a 19-page appendix on tank standards focusing on watertight construction, installation and testing (Engle, 2006).

Additionally, internal and external QA/QC at tank manufacturing facilities is critical if quality tanks are to be installed. Internal QA/QC is that implemented by the tank manufacturer. Anecdotally (and in some literature references), it is strongly suggested that many precast concrete tank manufacturers have inadequate QA/QC programs, although efforts by the National

Precast Concrete Association (NPCA) and a number of conscientious precasters are helping to facilitate better QA/QC in the industry as a whole.

Existing Tank Standards

A variety of existing tank standards were reviewed as a precursor to the detailed data gathering phase of this project in an effort to establish the landscape of current and historical practice and to begin to identify the more important studies influencing practice. The details of this review are only summarized here, and the reader is referred to Appendix A for additional detail.

This effort included several components:

- Review of a total of 57 fact sheets and/or packages of fact sheets
- Review of all identified primary unit industry standards
- Review of each state's primary unit regulations
- Review of most of the major texts, books and guidance manuals addressing decentralized wastewater management

In summary, the U.S. PHS studies and the associated PHS and EPA design manuals are still the dominant influence on today's regulations and standards. Industry standards and especially regulations were found to be generally conservative and slow to change.

State Regulations

The objectives of this evaluation were to:

- Establish the regulatory landscape for septic tanks and grease traps
- Determine the scope of existing regulations
- Assess specifics of existing regulations and their bases
- Identify unique or noteworthy regulations/programs to highlight

Most primary unit regulations appeared to share a common basis (U.S. PHS and EPA Design Manuals) but there was still great variety in their level of development and specifics. Based on a qualitative review of each state's regulations, less than half (18) had what would be considered "well-developed" primary unit regulations, although it should be noted that this doesn't necessarily mean that state has a poor septic tank program, as quality may be assured using non-regulatory means that weren't apparent in the review.

As indicated previously, state regulations are generally slow to change. For example, most states still do not require multiple compartment tanks, though many encourage and/or address them in regulations. Additionally, at least seven states allow vertical cylindrical tanks; the remainder address rectangular or horizontal cylindrical only. Seven states require effluent screens on all new tanks, while two additional states allow the use of an effluent screen to preclude a requirement to use some other septic tank enhancement. States addressing effluent screens typically reference NSF Standard 46 and/or maintain a list of state-approved products.

Grease traps are addressed in 21 states' decentralized regulations and the grease trap is typically specified to be similar to the septic tank design for that state, except with longer influent and effluent tees or baffles. Grease trap sizing is usually based on EPA or UPC formulas or on a straight hydraulic residence time basis (usually 24–48 hours).

Thirteen state regulations specifically indicate that prefabricated tanks must be approved by the state. Several states had what appeared to be very well-developed programs. For example, in Alabama, tank manufacturing facilities must be inspected twice per year, by regulation. The State of Massachusetts requires each manufacturer to implement a QA program in accordance with American Society for Testing and Materials (ASTM) C1227. Florida also defines its tank QA program in its regulations. Other states may have good QA programs not described in their rules. For example, the QA program for North Carolina has been reported in the literature (D'Amato and Devkota, 1997), but is not explicitly codified in their state regulations. Additionally, many local regulatory jurisdictions undoubtedly have tank QA programs. These were not investigated.

Most regulations specify that tanks must be watertight, but few contain QA requirements and measures to that end. Only three states' regulations require watertightness testing for all installed tanks, while one state requires testing in certain circumstances and two additional states describe watertightness testing, but do not require it.

Industry Standards

There are many redundant, sometimes conflicting industry standards, although there appears to be an increasing amount of cooperation and coordination among standards-setting efforts. This coordination is reflected in more common cross-referencing of one entity's standard by a different standard-setting entity. Industry standards are typically developed by consensus among a team or subcommittee of regulators, practitioners, contractors and other stakeholders. As such, many such standards are based more on harmony with existing practice (which is often a result of repetitive practice established by other regulatory or industry standards) than a fresh look at the science behind the practice.

Moves toward national performance-based codes could establish better consistency and, if implemented well, should facilitate the adoption of improved primary unit design features and operation practices. Additionally, increased availability of standardized education and training materials will contribute to a more nationally focused effort on this subject.

Table 6 provides a listing of the most relevant standards and codes directly addressing primary units in decentralized wastewater treatment systems.

Table 6. Broadly Applicable Standards and Codes Addressing Primary Treatment Units In Decentralized Wastewater Treatment Systems.

Organization	Standard Number	Standard Title
ASTM	C890-06	Standard Practice for Minimum Structural Design Loading for Monolithic or Sectional Precast Concrete Water and Wastewater Structures
ASTM	C1227-05	Standard Specification for Precast Concrete Septic Tanks
ASTM	C1644-06	Standard Specification for Resilient Connectors Between Reinforced Concrete On-Site Wastewater Tanks and Pipes
ASTM	C1613-06	Standard Specification for Precast Concrete Grease Interceptors
IAPMO	PS 001-2006	Prefabricated Septic Tanks
IAPMO	PS 080-2006	Grease Interceptors and Clarifiers
IAPMO	Appendix H, UPC	Recommended Procedures for Design, Construction and Installation of Commercial Kitchen Grease Interceptors
CSA	B66-05	Design, Material, and Manufacturing Requirements for Prefabricated Septic Tanks and Sewage Holding Tanks
NSF/ANSI	Standard 46	Evaluation of components and devices used in wastewater systems
ICC	IPC	International Plumbing Code
ICC		International Private Sewage Disposal Code
UL	UL 70	Septic Tanks, Bituminous-Coated Metal
AS-NZS[1]	1546, Part 1	Septic Tanks (1998)
CEN[2]	EN 12566, Part 1	Prefabricated Septic Tank
CEN[2]	EN 12566, Part 4	Septic Tanks Built *In Situ* from Prefabricated Kits

[1] Australia-New Zealand
[2] European Union

Research Recommendations

The preceding discussion revealed numerous unknowns about the performance of primary treatment units in decentralized systems; as such, there are a multitude of potential research questions that could be pursued. However, the answers to many such questions are likely to be inconsequential; that is, the results will not result in meaningful improvements in system performance. Furthermore, the resources required to answer some questions with confidence may not be worth the potential improvements in practice. There is thus a great need to consider all potential research objectives with respect to anticipated research costs versus potential benefits to overall decentralized system performance.

In light of the information presented above, twenty priority research areas and specific needs have been identified (listed in no particular order):

Overall System Performance

1. Evaluate the performance of alternative septic tank designs, relative to flow patterns, inlets and compartment orientation (e.g., meander, laminar flow, UASB septic tanks, Imhoff tanks, and others) and assess their applicability for various types of systems (e.g., nitrogen removal systems) and/or sites. Performance in this context should include the following assessments: suspended solids removal efficacy, hydraulic performance (effective HRT), sludge/scum accumulation rates, and potentially other measures depending on the application.

2. Evaluate grease trap performance (FOG removal efficacy) relative to sizing and design details, considering food service characteristics which are likely to be influential (note that WERF's FROG Project No. 03-CTS-16T is starting to do this).

3. Assess the field performance of effluent devices, via effluent sampling, with specific objectives of quantifying the effect of effluent screens and the impacts of variable screen design parameters on septic tank and grease trap performance (TSS, BOD_5, and FOG reductions). Population selection and sampling regimens will need to be robust enough to yield statistically meaningful results when considering the effect of outlet device only. (NSF International has expressed interest in such a project to inform Standard 46 for effluent filtration devices.)
4. Evaluate seasonal (e.g., temperature) and regional differences in primary unit function and optimize design, O&M recommendations accordingly.
5. Develop and calibrate CFD models and begin to use them, in concert with pilot and full-scale hydraulic/flow dispersion testing, to evaluate potential primary treatment design enhancements and to predict performance characteristics (note that WERF Project No. 03-CTS-16T is starting to do this for grease traps).
6. Evaluate viable energy efficiency and recovery strategies for primary treatment units. Biogas latrines—which combine toilet wastewater with household, agricultural and/or garden waste to generate biogas and fertilizer at the individual and community level—can be explored (Lens et al., 2001), as can pre-composting tanks as described by Gajurel et al. (2003). Additionally, collected oils and greases and oils can be converted into biofuels (Hake et al., 2006).
7. Further develop and evaluate tank intercompartment connectors and inlet devices to optimize existing, common septic tank designs.

<u>Performance as a Function of Wastewater Characteristics</u>

8. Assess performance of septic tanks serving cluster and specific non-residential facilities and perform an engineering study on scientifically defendable modifications (or not, if no practical enhancements exist) for primary treatment units serving these facilities.
9. Conduct NOWRA/WQA field survey of systems to determine impacts of water softeners on septic tank function and performance and conduct additional controlled studies as necessary to complete understanding of potential impacts and effects.
10. Develop and validate standard protocols for testing the safety/fate of household chemicals in septic tanks for acceptance of and adoption by the industry.
11. Generate a statistically sufficient set of paired influent and effluent data for in-service septic tanks in various geographies and under different design and operating conditions to better assess performance under real-world conditions. This effort could potentially leverage existing monitoring programs to create an ongoing database of performance data.
12. Evaluate fate, occurrence and treatability of priority pollutants and contaminants of emerging concern in primary treatment units. Some of this research is ongoing and the results would best be evaluated in the context of the overall decentralized system and receiving environment and associated environmental and human health risks.

Oversight and Operation

13. Develop a model tank construction and installation QA/QC program (with an emphasis on watertightness) that can be adopted and implemented by manufacturers, engineers and regulatory authorities. NPCA has already made significant strides in this regard for precast concrete tank manufacturers and NOWRA is continuing to work on a model performance code for watertightness. Construction, joint formation (or elimination), and installation practices could be optimized to assure watertightness.
14. Improve understanding of the evolution/maturation of septic tank biological function and associated impacts on design and operation. Reconsider solids pumping protocols in light of a more holistic understanding of septic tank biological function.
15. Develop and validate standard methods for measuring and monitoring septic tank performance (e.g., influent and effluent sample collection and analysis) for use in future studies and in routine system monitoring efforts.
16. Develop and validate standard methods (manual and automatic) for measuring sludge and scum accumulations in septic tanks for use in future studies.
17. Develop methods to better assess septic tank biology and septic tank function in the field and improve laboratory/microscopic techniques for diagnosing upsets.
18. Provide a mechanism for industry to evaluate the actual implications and effectiveness of tank additives. This could include developing and validating standard protocols for testing the effectiveness of septic tank additives for adoption by the industry. A number of different testing options may be necessary given the variation in performance claims for different products.
19. Develop and establish the validity of field sites for testing septic tanks and grease traps. Groves et al. (2005) conducted a study comparing controlled field site data with that of field data for NSF International Schedule 40 testing. Their statistical analysis showed that the variability associated with the controlled site data was significantly less than that associated with the field sites and, as such, one dataset cannot be used to predict the other.
20. Determine whether there is a correlation between state regulations on construction, installation and inspection and failure rates.

Types of Studies

Three main types of studies are represented in the existing body of research: laboratory-scale studies, controlled pilot testing and field surveys. Each is appropriate in certain circumstances and of limited utility in others.

Laboratory-Scale Studies are generally the least costly and easiest to control; however, they suffer from lack of realism and associated scaling issues. These studies are useful for answering fundamental or broad research questions that either can be subsequently field-tested or are sufficient in their own right. Bench-scale toxicity testing for the screening of additives or chemicals, for example, may be sufficient for assessing biological effects without confirmatory field testing, provided that the methods have been predetermined to be robust. Note that Peeples and Mancl (1998) describe the development of a laboratory-scale septic tank which maintains sludge and scum layers and provides a relatively consistent effluent with realistic variability in quality.

Controlled Pilot Studies are the most common types of tests that have been employed in the field to date. These studies usually utilize full sized treatment units, but with a controlled influent which potentially simplifies the experimental design and statistical analysis of results. However, the influent is often unrealistic from both a quality and hydraulic standpoint limiting the application of results to the extremely variable conditions often experienced in the field. Controlled pilots should continue to be important elements of testing, although standard procedures need to be better established so that the results of such studies and their limitations can be better understood and used to inform the application of results to practice.

Field Surveys can provide the most powerful information on treatment unit performance, as real-world variability is captured in the results. Groves et al. (2005) determined that for aerobic treatment units, the source of variability between residential sites is generally greater than that within a residential site. Because of this variability—which can be extreme—statistically robust experimental designs can be difficult to develop and very expensive to carry-out. Additionally, lack of control over experimental conditions can complicate statistical analysis of the results. Nevertheless, field surveys should have great importance in the assessment of primary unit performance.

Common Study Design Limitations

Per this extensive literature review, several categorical limitations or potential limitations of studies can be generalized, as follows:

Variable Selection and Statistical Power: Many studies have been conducted with too many variables under consideration, making it impossible to isolate the effects of individual variables on performance and thus yielding little useful information. Likewise, many reported studies do not clearly address the statistical significance of their results. Future studies need to be more realistic and set clear research objectives that can be accomplished with available resources. Additionally, the results of such well-designed research should be published in respected, peer-reviewed journals; there is currently a paucity of such publications in the literature.

Sludge Seeding and Biological Maturation: Some controlled pilot studies have been conducted by adding sewage sludge (often from a municipal wastewater treatment plant) to a septic tank to simulate natural septage accumulation. This may be appropriate from the standpoint of simulating performance of a septic tank as a settling unit, but is unrealistic in its consideration of the extremely important biological functions and effects inherent to septic tanks. Even septic tanks in operation for months may not adequately simulate the performance of a septic tank with more mature biology.

Paired Influent/Effluent Data: Heavy reliance on STE data as the primary performance measure is inadequate in many situations. While average effluent quality is an important performance criterion, the variability in inlet and effluent quality should also be considered (Otis and Boyle, 1976). Because influent characteristics are such a fundamental variable inherent to virtually all controlled pilot and field survey experimental designs, paired influent and effluent data may be needed to adequately characterize performance, depending on the specific objectives of the research in question.

Existing Data

During project planning, it was assumed that there was an existing untapped pool of data that could be used to inform primary unit practice in a meaningful way. However, it appears that most of the truly useful data in the field has been published in some forum (primarily conference proceedings). No useful, unpublished datasets were encountered despite making numerous efforts to identify such datasets from leading researchers and practitioners and from national decentralized wastewater demonstration projects. Monitoring data (e.g., regulatory) undoubtedly exists in some quantity and may be useful but is not currently accessible in a practical way.

What information that is available in the decentralized wastewater arena could be better managed and distributed. The industry as a whole should attempt to develop tools that will allow future (and perhaps existing) monitoring data to be banked collectively. Such a system would need to be simple enough to encourage its use by operators, regulators, researchers and others, but would also have to be detailed enough to fully establish necessary QA information about the datasets entered.

Future studies can use the results of this project to identify useful existing data; however, it is recommended that the focus be on using available resources to conduct original research, focusing on statistically robust experimental designs.

Conclusions

Septic tank and grease traps are exceptionally efficient primary treatment units of critical importance in decentralized wastewater systems. Despite an appearance of elegant simplicity, the function of such units is quite complex and their performance is likely affected by numerous factors, of which relatively few have been assessed in a level detail sufficient to definitively influence design or operation. Additionally, much of the past work has yielded datasets with insufficient statistical power to inform design. Future work should be focused and set clear research objectives that can be accomplished with available resources, and such well-designed research results should be published in respected, peer-reviewed journals.

Additional efforts to understand and ultimately improve primary unit performance are likely warranted. However, full understanding of every factor that potentially influences performance is an unachievable goal whose results would be of limited use relative to the effort expended. As such, future efforts should be directed toward answering outstanding questions whose results are likely to yield real improvements in design, operation, and associated areas and within the context of overall decentralized wastewater system performance. A comprehensive research strategy and agenda should be developed to determine which research questions are worthy of further study relative to costs/benefits and risk management.

This research digest and associated bibliographic database will assist future researchers, especially, in their efforts. Other practitioners (designers, engineers, regulators) will find the products useful as tools for investigating specific primary treatment issues they encounter in their practice. Future efforts should build upon this one by continuing to update the products and tools and developing similar tools for other aspects of decentralized wastewater management. As future research is conducted and presented, it should be added to the database and integrated into the digest.

Additional Resources

Several interrelated products associated with the research digest are available. The following products are available at www.werf.org and www.ndwrcdp.org:

- **Research Digest**: Factors Affecting the Performance of Primary Treatment in Decentralized Wastewater Systems
  - Research Digest Appendices
- **Bibliographic Database** of Research and Data on Performance of Primary Treatment Units in Decentralized Wastewater Systems
  - Bibliographic Database Readme
- **Fact Sheets and Communications Tools**
  - Technical Guide: Primary Treatment in Decentralized Wastewater Systems (succinct guidance document)
  - Primary Treatment in Decentralized Wastewater Systems: Research Considerations (fact sheet)
  - Primary Treatment in Decentralized Wastewater Systems: Policy Points (fact sheet)

APPENDIX A

EXISTING TANK STANDARDS

A variety of existing tank standards were reviewed as a precursor to the detailed data gathering phase of this project in an effort to establish the landscape of current and historical practice and to begin to identify the more important studies influencing practice.

This effort included several components:

- Review of a total of 57 fact sheets and/or packages of fact sheets
- Review of all identified primary unit industry standards
- Review of each state's primary unit regulations
- Review of most of the major texts, books and guidance manuals addressing decentralized wastewater management

In general, the U.S. PHS studies and the PHS and U.S. EPA design manuals are still the dominant influence on today's regulations and standards. Industry standards and especially regulations were found to be generally conservative and slow to change. Additionally, there are many redundant, sometimes conflicting industry standards, but there appears to be an increasing amount of cooperation and coordination among standards-setting efforts. Moves toward national performance-based codes could establish better consistency and, if implemented well, should facilitate the implementation of improved primary unit design and operation practices.

Industry Standards

The landscape for standards and codes that address primary treatment units in decentralized systems is somewhat cluttered with often duplicative and inconsistent standards. Some standards are commonly cited, while others appear to have more limited relevance or use. The three most commonly cited septic tank design/construction standards are those developed by ASTM International, the International Association of Plumbing and Mechanical Officials (IAPMO), and the Canadian Standards Association (CSA). There is a reasonable amount of consistency between these standards and there appears to be increasing effort to coordinate the future development of these standards. NSF International appears to have the only accepted standard for septic tank effluent filtration devices. It appears widely-used as the standard for these products.

The landscape for large outdoor grease interceptor standards appears to be even more fragmented. Most true standards for these units are contained in the two major plumbing codes, the Uniform Plumbing Code (UPC) and the International Plumbing Code. There are more standards applicable to small indoor grease interceptors, although these are of much less significance to decentralized systems.

Table A-1 provides a brief summary of the most relevant standards and codes directly addressing primary units in decentralized systems. A discussion of the major relevant standards and codes is provided below, sectioned by issuing organization. Note that much of the information provided below is taken from websites and other documentation provided by the issuer. Links are provided, as applicable. Where standards are described below, specifics are sometimes left out so as to not compromise the proprietary nature of the standards.

ASTM International

ASTM International, formerly known as the American Society for Testing and Materials (ASTM), "is one of the largest voluntary standards development organizations in the world – a trusted source for technical standards for materials, products, systems, and services" (www.astm.org).

With regards to septic tank and grease trap design and construction, ASTM has focused on tanks constructed of precast concrete. Several ASTM standards have been adopted into state regulations and engineering specifications for the design, construction, installation and testing of tanks constructed of other materials, including fiberglass and polyethylene, but no standards specific to septic and grease traps utilizing these materials currently exist. The focus of this discussion will be on those standards specifically pertaining to septic tanks and grease traps.

Concrete septic tank and grease trap standards are currently under the purview of ASTM C27.30. ASTM Committee C27, Precast Concrete, has jurisdiction of 24 standards. C27 has 6 technical subcommittees, including C27.30, Water and Wastewater Containers.

IAPMO

The International Association of Plumbing and Mechanical Officials (IAPMO; http://www.iapmo.org/) is the largest plumbing code writing body in the world. The IAPMO Standards Department represents IAPMO in many standards-writing organizations throughout North America, including the American National Standards Institute (ANSI), the ASTM, the American Society of Mechanical Engineers (ASME), the American Society of Sanitary Engineering (ASSE), the CSA, and the Recreational Vehicle Industry Association (RVIA). The IAPMO Standards Department is also involved in its own standards work. IAPMO publishes over 90 current IAPMO property standards. These are standards which have been developed through IAPMO's standards development process. Many of IAPMO property standards have been developed into nationally recognized standards in conjunction with organizations such as ANSI, ASME, ASSE, and ASTM. With the help of industry, IAPMO also writes and revises installation standards, which are guidelines for proper installation of plumbing products.

IAPMO has two primary standards of interest to primary treatment units in decentralized wastewater systems (http://www.iapmostandards.org/). These include *IAPMO PS 001-2006 – Prefabricated Septic Tanks* and *IAPMO PS 080-2006 – Grease Interceptors and Clarifiers* (which references the Uniform Plumbing Code, also an IAPMO product).

Table A-1. Broadly Applicable, Cited Standards and Codes Specifically Addressing Primary Treatment Units in Decentralized Systems.

Organization	Standard Number	Standard Title	Scope/Description
ASTM	C890-06	Standard Practice for Minimum Structural Design Loading for Monolithic or Sectional Precast Concrete Water and Wastewater Structures	This practice describes the minimum loads to be applied when designing monolithic or sectional precast concrete water and wastewater structures with the exception of concrete pipe, box culverts, utility structures, and material covered in Specification C 478.
ASTM	C1227-07	Standard Specification for Precast Concrete Septic Tanks	This specification covers design requirements manufacturing practices, and performance requirements for monolithic or sectional precast concrete septic tanks.
ASTM	C1644-06	Standard Specification for Resilient Connectors Between Reinforced Concrete On-Site Wastewater Tanks and Pipes	This specification covers the minimum performance and material requirements for resilient connectors used for connections between reinforced concrete tanks used for septic effluent treatment/detention, including those referenced in Specifications C 913 and C 1227.
ASTM	C1613-06	Standard Specification for Precast Concrete Grease Interceptor Tanks	This specification covers design requirements, manufacturing practices, and performance requirements for monolithic or sectional precast concrete grease interceptor tanks. This specification describes precast concrete tanks installed to separate fats, oils, grease soap scum, and other typical kitchen wastes associated with the food service industry.
IAPMO	PS 001-2006	Prefabricated Septic Tanks	The purpose of this standard is to establish an acceptable quality standard for prefabricated septic tanks of concrete, fiberglass reinforced plastic or polyethylene intended for domestic sewage disposal systems. It shall serve as a guide for producers, distributors, architects, engineers, contractors, installers, inspectors and users, and to promote a better understanding regarding materials, manufacture, and installation. The provisions of this standard are not intended to prevent the use of any alternate material or method of construction, provided any such alternate meets the intent of this standard.
IAPMO	PS 080-2006	Grease Interceptors and Clarifiers	The purpose of this standard is to establish specifications regarding the construction of grease interceptors and clarifiers. Grease interceptors are sized and specified under Appendix H of the Uniform Plumbing Code. Its purpose is to serve as a guide for producers, distributors, architects, engineers, contractors, installers, inspectors and users; to promote understanding regarding design, materials, and the installation; and to provide for identifying grease interceptors and clarifiers that conform to this standard.
IAPMO	Appendix H, UPC (2003)	Recommended Procedures for Design, Construction and Installation of Commercial Kitchen Grease Interceptors	Provisions apply to the design, construction, installation, and testing of commercial kitchen grease interceptors.

Organization	Standard Number	Standard Title	Scope/Description
CSA	B66-05	Design, Material, and Manufacturing Requirements for Prefabricated Septic Tanks and Sewage Holding Tanks	This Standard specifies minimum design and material requirements as well as manufacturing practices and markings for prefabricated septic tanks, sewage holding tanks, and effluent chambers of steel, concrete, fiberglass-reinforced plastic, or polyethylene construction that are designed to handle sewage or sewage effluent.
NSF/ANSI	Standard 46	Evaluation of components and devices used in wastewater systems	Standard 46, relating to components of wastewater treatment systems, applies to any wastewater treatment product not covered by Standards 40 and 41. Standard 46 currently includes performance evaluations for grinder pumps, septic tank effluent screens, chlorination devices, and UV disinfection devices.
ICC	IPC	International Plumbing Code	Contains comprehensive minimum regulations for plumbing facilities in terms of both performance and prescriptive objectives, and provides for the acceptance of new and innovative products, materials and systems.
ICC		International Private Sewage Disposal Code	Companion to the IPC which includes provisions for design, installation, and inspection of private sewage disposal systems, and provides flexibility in the development of safe and sanitary systems.
UL	UL 70	Septic Tanks, Bituminous-Coated Metal	The standard covers single and multiple compartment, bituminous-coated metal septic tanks for use in residential sewage systems.

IAPMO PS 001-2006, Prefabricated Septic Tanks, was originally ratified in 1958, making it the oldest specific septic tank standard still in use in the U.S., and has since been revised eleven times, most recently in 2006. This 16 page standard references a number of ASTM standards, including C1227. The stated purpose of this standard is to establish an acceptable quality standard for prefabricated septic tanks of concrete, fiberglass reinforced plastic or polyethylene intended for domestic sewage disposal systems. According to PS 001-2006, septic tank design shall be such as to produce a clarified effluent consistent with accepted practice and shall provide adequate space for sludge and scum accumulations.

IAPMO PS 080-2006, Grease Interceptors and Clarifiers, was originally adopted in 1995 and has since been revised four times, most recently in 2006. The stated purpose of this standard is to establish specifications regarding the construction of grease interceptors and clarifiers. According to PS 080-2006, grease interceptors shall be designed to remove grease from effluent and to retain grease until accumulations can be removed by pumping the interceptor. It is noted that grease interceptors are sized and specified under Appendix H of the Uniform Plumbing Code.

Appendix H, Uniform Plumbing Code (UPC), Recommended Procedures for Design, Construction and Installation of Commercial Kitchen Grease Interceptors (2003), is frequently cited with respect to sizing and design of large outdoor grease interceptors. The UPC is governed by IAPMO. A number of modifications have been approved for the 2006 edition, including changes in the grease interceptor sizing criteria. The new formulas are based on the number of drainage fixture units (DFUs), rather than peak number of meals served. In addition, the code will more clearly differentiate between large outdoor interceptors and the indoor, under-the-sink type, by referring to them as gravity-type and hydromechanical interceptors, respectively. Finally, the new sizing criteria will be moved into Chapter 10 and removed from Appendix H, which will be eliminated (http://www.iapmostandards.org/).

Canadian Standards Association (CSA)

The CSA (www.csa.ca) is a not-for-profit membership-based association serving business, industry, government and consumers in Canada and the global marketplace. The CSA works in Canada and around the world to develop standards. Their stated goals are enhancing public safety and health, advancing the quality of life, helping to preserve the environment and facilitating trade.

The CSA additionally provides certification and testing under the name CSA International. CSA International provides certification services for manufacturers who, under license from CSA, wish to use the appropriately registered CSA Marks on certain products of their manufacture to indicate conformity with CSA Standards.

CSA B66-05, Design, material, and manufacturing requirements for prefabricated septic tanks and sewage holding tanks, was last published on April 1, 2005. This is the sixth edition of CSA B66 and supersedes the previous editions published in 2000, 1990, 1985, 1979, and 1975 under the title Prefabricated Septic Tanks and Sewage Holding Tanks.

The primary scope of CSA B66-05 is stated as follows: "This Standard specifies minimum design and material requirements as well as manufacturing practices and markings for prefabricated septic tanks, sewage holding tanks, and effluent chambers of steel, concrete, fiberglass-reinforced plastic, or polyethylene construction that are designed to handle sewage or sewage effluent."

The standard is comprehensive in that it includes criteria for nearly all elements of septic tank design, construction and testing. It does not, however, include specifications for installation of septic tanks. The 38-page standard includes sections on general requirements, steel tanks, concrete tanks, fiberglass-reinforced polyester tanks, polyethylene tanks, testing, and markings.

Much like ASTM C1227, B66-05 references multiple other CSA standards, mostly those pertaining to materials used for tank construction. It also references a number of ASTM standards (but not ASTM C1227) and ULC standards (Underwriters' Laboratories of Canada).

B66-05 includes figures showing typical tank layouts. Gravity outlet tanks are termed "trickle-type septic tanks", as opposed to "siphon-type septic tanks" which include an effluent chamber furnished with a siphon discharge. For trickle-type septic tanks, single or dual compartment models are described.

NPCA

The National Precast Concrete Association (NPCA) is an international trade association, which represents manufacturers of plant produced precast concrete products and companies that provide the equipment, supplies and services to make these products. NPCA also provides technical and production information through 13 product committees, which consist of members who concentrate on specific product lines within the precast industry (www.precast.org).

The Second Edition of NPCA's *Best Practices Manual: Precast Concrete On-site Wastewater Tanks* was released in October 2005. This is not a standard, per se, but rather a guide for manufacturers, inspectors, engineers, contractors and the like. The 30 page manual is presented in an appealing format, with extensive use of figures, photographs and easy-to-read typesetting. The content is also very thorough, covering most aspects of concrete tank construction in a practical manner, while providing references for applicable standards or specific design criteria to be used. NPCA has a companion video entitled, "Producing Watertight Concrete Septic Tanks."

For design, all three major septic tank standards are referenced (ASTM, IAPMO, and CSA), and no specifics are given. American Concrete Institute (ACI) standards are frequently referenced for concrete practices. However, the focus of this manual is on quality control with respect to construction, installation and testing. The sections of the manual include: structural design; materials; concrete mixture proportioning; lifting inserts; coatings; production practices; reinforcement; casting concrete; curing procedures; post-pour operations; finishing and repairing concrete; seals, fittings and joints; tank installation; and testing, which includes methods for concrete compressive strength testing, vacuum testing and water testing.

Finally, NPCA also has a Quality Control Manual for Precast Concrete Plants upon which a Plant Certification Program to assist precasters with quality control is based. The Plant Certification Program includes a relatively new accreditation program specific to precasters of onsite wastewater equipment (http://www.precast.org/owap/index.htm)

Design Considerations and Discussion of Large Outdoor Grease Interceptors (http://www.precast.org/technical/Grease_Interceptor_Design_s.pdf) is a short manual from NPCA that strives to provide clarification on the field-proven performances of available animal and vegetable fats oils and grease (AVFOG) removal technologies due to the lack of definitive literature on the design, construction and operation of large, outdoor grease interceptors. The premise is that small indoor grease traps are inadequate and that large outdoor grease interceptors are necessary. The document is more geared toward public sewer systems, but the take-home message is only magnified for more sensitive decentralized systems.

American Concrete Institute (ACI)

ACI (http://www.concrete.org/general/home.asp) has published a number of specifications that are routinely cited in construction standards for concrete tanks. Most commonly cited standards include the following overarching standards; however, ACI has other applicable codes for more specific issues.

318-05/318R-05: Building Code Requirements for Structural Concrete and Commentary.

350/350R-01: Code Requirements for Environmental Engineering Concrete Structures and Commentary.

NSF International

NSF International was founded in 1944 as the National Sanitation Foundation. As a not-for-profit, non-governmental organization, NSF is a leading global provider of public health and safety risk management solutions, providing product certification and safety audits for the food and water industries (www.nsf.org).

NSF/ANSI Standard 46, Evaluation of components and devices used in wastewater systems, currently includes performance evaluations for grinder pumps, septic tank effluent screens, chlorination devices, and UV disinfection devices. Section 10 of the standard, *Filtration devices for residential gravity flow septic tank systems,* is the part of the standard specific to effluent screens. Standard 46 only applies for flow rates between 400 and 1,500 gallons per day.

In Standard 46, effluent screens are defined as gravity-flow devices designed to enhance the retention of solids in a residential septic tank system. Certification for meeting the standard is achieved through a series of tests conducted by NSF at their test facility in Chelsea, Michigan. For effluent screens, the tests include: a flow test for clean screens, flow test for partially clogged screens, a structural integrity test, a filtration efficacy test, and a by-pass protection test.

Manufacturers with approved effluent screens under Standard 46 can be found at http://www.nsf.org/Certified/Wastewater/.

International Code Council (ICC)

The International Code Council (http://www.iccsafe.org/), a membership association dedicated to building safety and fire prevention, develops the codes used to construct residential and commercial buildings, including homes and schools.

The 2006 *International Plumbing Code* (IPC) provides comprehensive minimum regulations for plumbing facilities in terms of both performance and prescriptive objectives, and provides for the acceptance of new and innovative products, materials and systems. The IPC appears to defer to ASME standards (listed below) for grease control device standards in Chapter 10 of the Code. The standards appear to be focused toward the small indoor mechanical interceptors.

The 2006 *International Private Sewage Disposal Code* is a companion to the IPC which "includes provisions for design, installation, and inspection of private sewage disposal systems, and provides flexibility in the development of safe and sanitary systems". Chapter 8 is entitled Tanks, and includes several pages on septic tanks and their maintenance.

ASME International

ASME (formerly the American Society of Mechanical Engineers) is a 120,000 member professional organization focused on technical, educational and research issues of the engineering and technology community. ASME conducts one of the world's largest technical publishing operations, holds numerous technical conferences worldwide, and offers hundreds of professional development courses each year. ASME sets internationally recognized industrial and manufacturing codes and standards that enhance public safety (www.asme.org).

ASME lists two standards related to grease removal systems, but both appear to be targeted to the small indoor mechanical products. Descriptions from ASME are provided below.

A112.14.3 - 2000 Grease Interceptors
No. of pages: 27
Reaffirmed Date: 2004

Description: This Standard covers general product requirements as well as the performance criteria for the testing and rating of grease interceptors, whose rated flows are 100 gpm (380 L/m) or less.

A112.14.4 - 2001 Grease Removal Devices
No. of pages: 14

Description: This Standard establishes requirements for grease interceptors that are equipped with automatic grease removal devices (GRD). It includes testing requirements and performance criteria designed to ensure conformance to this Standard. Such devices are designed for the purpose of automatically removing free-floating grease, fats, and oils from sanitary discharges without intervention from the user except for maintenance. Semiautomatic devices are not addressed in this Standard.

Underwriters Laboratories, Inc.

Underwriters Laboratories Inc. (UL) is an independent, not-for-profit product-safety testing and certification organization (http://www.ul.com/). Standard UL 70, Septic Tanks, Bituminous-Coated Metal (http://ulstandardsinfonet.ul.com/catalog/stdstd.html), dated August 29, 2001, covers single and multiple compartment, bituminous-coated metal septic tanks for use in residential sewage systems. This standard was not reviewed; however, it contains sections on materials and construction, watertightness, capacity, single and multiple compartment tanks (sizes and dimensions, baffles and fittings, access), coatings and markings.

NOWRA Model Performance Code

NOWRA is the largest organization within the U.S. dedicated solely to educating and representing members within the onsite and decentralized industry. The NOWRA White Paper, authored by the leadership of the NOWRA Model Performance Code Committee, presents the concept of a national Model Performance Code. It was approved by the NOWRA Board, June 12, 2004 and formally adopted September 28, 2004. The paper presents the advantages of using performance over prescription codes in making decisions on the use and location of onsite or cluster wastewater treatment systems and addresses (www.nowra.org).

The Septic Tank Task Group is developing a NOWRA performance classification system and protocol for septic tanks relative to installed system watertightness. The interest in "truly watertight" tanks is strong on the primary code committee and this task group.

State Regulations

The objectives of this evaluation were to:

- Establish the regulatory landscape for septic tanks and grease traps
- Determine the scope of existing regulations
- Assess specifics of existing regulations and their bases
- Identify unique or noteworthy regulations/programs to highlight

All forty-eight states having statewide onsite regulations (California and Michigan are currently regulated at the local level) were collected from the National Small Flows Clearinghouse (NSFC) database (http://www.nesc.wvu.edu/nsfc/nsfc_regulations.htm).

Sections dealing with these primary units were isolated and reviewed to glean notable information from them. This was done for each state and the results were summarized in shorthand form in an excel spreadsheet file, Table A-2, containing the following fields (columns):

- State
- Development of Regs. – a brief qualitative assessment of the state of the development of that state's septic tank regulation (generally: none, minimal, partially-developed, well-developed, extensive)
- Contact – contact name for potential follow-up
- Sizing – information on septic tank sizing requirements
- Compartmentation – information on compartmentation requirements, including whether multiple compartments are mandated, the sizing of compartments, openings in compartments, use of tanks-in-series, etc.
- Dimensions – notable standards specified for interior tank dimensions, including range of depths, length-to-width-to-depth ratios, allowable tank shapes, etc.
- Inlet – style and submergence depth of inlet structure, also whether pipe penetrations need to meet a standard
- Outlet – style and submergence depth of outlet structure, use of effluent screens and/or gas baffles, also whether pipe penetrations need to meet a standard
- Access – style, location of access openings and relation of opening/lid to finished grade
- Construction – construction materials mentioned (allowed/prohibited), amount of construction information on each, and referenced construction/material standards, if any
- Grease Traps – whether grease traps are addressed or required, sizing and design information
- Preapproval – whether a tank or manufacturer approval program is laid out in the regulations
- Other/Comments – any other information, including whether the regulation includes installation procedures, abandonment procedures, mandatory watertightness testing, information about additives, and the like

Table A-2. Summary of State Regulatior s on Primary Treatment Units in Decentralized Wastewater Systems.

State	Development of Regs.	Sizing	Compartmentation	Dimensions	Inlet	Outlet	Access	Construction	Grease traps	Preapproval	Other/Comments
Alabama	well-developed	2 day min HRT, by bedroom, 750 min	2, first at 2/3 length, 3 types of baffle walls. Two-in-series allowed	w = 3' min, l = 1.5-2Xw, d = 3-6'	ell below liquid	tee 6" out and 18" below water	to grade for non-residential	concrete, fiberglass, plastic	for food svc facilities; designed per EPA; 48 hr min HRT	yes. Min 2 yard inspections/year!	impressive tank quality assurance program!
Alaska	partially-developed	1000-gal up to 3 bdrm, then 250 per bedrm; EPA equation for larger tanks	most design criteria is apparently in AK's plumbing code (app. K)				apparently not necessarily to grade; emphasize security of covers				most design criteria is apparently in AK's plumbing code (app. K)
Arizona	minimal										apparently all septic system have to be designed by an engineer. Application must include septic tank capacity.
Arkansas	extensive	2 day min HRT, by bedroom, 750 min. 1000 min for slow perking soils	unclear whether required. All compartments at least 250 gal, inlet compt at lesat 350 gal.	d = 3-6 '; l at least 36"; l = 2-3Xw for rectangular tanks	tee, 6" below liquid level. Baffle may be used in addition.	tee, 40% below liquid level (for rectangular), 35% for cylindrical. Baffle may be used in addition. Deflection devices recommended, not required	over each compartment, but only brought close to grade	precast concrete, fiberglass (IAPMO std)	for food svc facilities. References AR plumbing code. Gives table of flow rates for fixtures and size based on 4 min. retention time.	yes.	rationale wording is very clear
California	no state regs										CA is in the process of developing statewide regulations
Colorado	partially-developed	30 hr HRT. 750 min for 2 bd, with 250+ for each bd	2, or 2 in series. First compt, at least 50% total. Opening 14" below outlet invert	d = 30-60" or max of tank length		tee at least 14" below liquid	no more than 8" below grade				also has a simple tank decommissioning regulation.
Connecticut	extensive	1000-gal up to 3 bdrm, then 250 per bedrm; 24 hour HRT for larger tanks, 2 hour HRT at peak flow. If septic tank fed by a pump, add 50%. Add capacity for large bathtubs.	2, with 2/3 capacity in first. 2 in series allowed	ASTM C1227 with exceptions. D = 36" min.	baffle extended 8-18" below liquid, gasketted connection	baffle extendeing 10" to 40% below liquid. Effluent filter required for new tanks. Gasketted connection	over inlet and outlet, no more than 12" below grade. Tanks over 15' long shall have three.	concrete	24 hour HRT, one tank for 500-2000 gpd, 2 tanks for > 2000 gpd system. In/out baffles to 12" above tank bottom. At grade access over both.		includes watertightness testing spec and abandonment procedures. Have effluent filter approval program; approved filters in Appendix B. Unclear if there is preapproval of prefab tank designs or a tank yard QA program. Otherwise, very modern, well-thought out rules.
Delaware	well-developed	1000 gal up to 500 gpd. 500-15000 gpd = 1.5Q (1500 min); >15,000 gpd, case-by-case	rectangular, 2-compt, 2/3 capacity in first. Baffle can have tee or openings, no more than 40% into liquid depth.		tee or baffle of PVC or concrete, extending 12" to 40% into liquid. Concrete/bentonite or rubber gasket connection.	tee or baffle of PVC or concrete, extending 12" to 40% into liquid. Concrete/bentonite or rubber gasket connection. Require effluent filter approved by dept.	above-grade access to each compartment.	precast, cast-in-place concr.	min 1000 gal, multi-compt. Inlet tee/baffle 24" min into liquid, outlet tee/baffle 8" above tank bottom. At-grade access. Baffle wall tee 12" from tank bottom		includes an abandonment reg. Good installation reg. Watertightness testing required for built in place tanks. Have a design for a septic tank lift station.
Florida	extensive	table listing required septic tank capacities up to 5000 gpd.	multiple compartments or tanks in series - first at least 2/3, second at least 1/5. tanks in series over 3500 gpd - first tank is 1/2 - 4/5 total vol. Connection by slot or tee 30-40% of liquid depth.	d = 42-84";	may be a tee, etc., but not required. If tee, etc., no more than 33% of liquid depth. Connection by ASTM C923	effluent filter approved by state. Tee or baffle 30-40% into liquid depth. Solids deflection device if no effluent filter.Connection by ASTM C923.	access within 8" for inlet and outlet. Extensive specs on access.	concrete (extensive regs.), fiberglass, PE (IAPMO reference)	single or multiple compartment. Sizing per EPA equation?	yes, extensive program described including four different tank designs based on soil saturation and cover.	includes sizing for graywater septic tank. Includes abandonment reg. Very extensive tank QA program is described. Tank installation regulation.

State	Development of Regs.	Sizing	Compartmentation	Dimensions	Inlet	Outlet	Access	Construction	Grease traps	Preapproval	Other/Comments
Georgia	partially-developed	24 hour HRT min., 1000 gal up to 4 bd, 250+ for each adl bd. 50% + for garbage grinder.	2 compartment, with first at 2/3 or more. Tanks in series allowed if first is at least 1000 gal and at least as large as second		ASTM 3034 rated	ASTM 3034 rated. Effluent filter required			approved in accordance with Manual.	yes	refers to On-Site Sewage Management Systems Manual for approved effluent filters. Basic installation spec.
Hawaii	minimal	750 (2), 900 (3), 1000 (4), 1250 (5)						requires coating entire interior of concrete tanks			references 10-state standards for septic tanks (section 20 of chapter 20, excluding 20.102 and 20.202). Most of reg is installation requirements.
Idaho	partially-developed	900 (1-2 bd), 1000 (3-4 bd) + 250 for adl. Other flows V = 2Q, 750 min.	unclear whether required. Allows tanks in series if initial tank or compartment has 1/2 to 2/3 capacity	all interior dimensions at least 2 feet. D = 2.5-5 feet.	vented tee or baffle	vented tee or baffle, extending 40% depth for rectangular, 35% for cylindrical	over each compartment, inspection ports over inlet/outlet. w/in 18" finished grade	prohibits steel tanks. Includes precast concrete specs.		yes	includes some installation and abandonment procedures. Includes section for large systems: L:W ratio at least 3:1; inlet to allow for even distribution across tank width; W:D ratio between 1:1 and 2.25:1
Illinois	well-developed	per table. For other facilities 750 gal up to 500 gpd, 1.5Q for over 500 gpd.	single compartment requires gas deflection baffle. Multiple compartments or tanks required for flows over 1,350 gpd, with first 1/2-2/3 total volume. Opening at 40% depth. Multiple compartment tanks require gas deflection baffle.	d = 42-72";	cast-in watertight penetrations; inlet baffle within 12" of inlet, extends 6" into liquid	cast-in watertight penetrations; outlet baffle within 6" of outlet and extends 40% of liquid depth	over inlet and outlet, within 12" of grade	concrete, built in place, fiberglass, plastics		yes	solids retention tank required between source and septic tank when garbage grinder is used. Includes installation and abandonment procedures.
Indiana	partially-developed	750 min for 2 bd, 250 for each adl. Up to 5; then 150 per bd.		d = 30" - 6.5'	baffle or tee extending at least 6" deep, 8-12" from wall	baffle or tee extending 40% into liquid depth 4-6" from wall; gas deflection baffle	8" diameter at-grade for each tank or compartment	minimal requirements for concrete, cast in place and block tanks			allows drainholes, but must be plugged prior to use.
Iowa	well-developed	1000 min for 3 bd, 250 for each adl bd. Add 250 for high volume fixture, garb disposal, water softener	two, with first 1/2-2/3 total; second 1/2 - 1/3. two tanks in series allowed; slot of holes or tee into liquid 1/3 depth	d = 40" - 6.5'; l = at least 5' and at least 1.5x width. W = 2' min. Vertical cylindrical tank diameter at least 5'; 36" min from inlet to outlet of each compartment	tee between 8" and 20% liquid depth. Requires gasketted connections cast in place.	tee between 10" and 25% into liquid depth. Requires gasketted connections cast in place.	18" over each compt, or 24" over center wall (w/6" inspection ports over inlet and outlet), brought to within 6" grade.	concrete or plastic (no metal)			large and non-domestic have separate design criteria in appendix a/o from authority.
Kansas	well-developed	1000 min up to 3 bd, then 1200 (4), 1500 (5). V = 3Q required for garbage grinders	multiple compartments may be used. Opening at 35%d, and at least 12 square inches	d= 2-6.5'. L:W = 1.5-4. l = 6' min. w = 4' min.	tee 8" - 20%	tee 35%d (30% for cylindrical)	6" inspection ports to grade over inlet/outlet. Location riser for middle access port to finished grade	concrete. Mention of fiberglass and PE			specifies design for 20 year life. Installation spec, including watertightness testing (not required). Lots of reg is written as reccomendations, not requirements. Nice formatting.
Kentucky	partially-developed	1000 min for 3 bd, 250 for each adl bd. Add 250 for garb disposal. V = 1.5Q for other facilities							500 gal up to 6000 gpd, 1000 gal over 6000 gpd		minimal requirements, but appear to be based on soil and/or pretreatment system type.

State	Development of Regs.	Sizing	Compartmentation	Dimensions	Inlet	Outlet	Access	Construction	Grease traps	Preapproval	Other/Comments
Louisiana	partially-developed	V = 2.5Q, 500 gal min.	multi encouraged, but not required. Max of 3 compartments. Min 250 gal compartment. In-series also encouraged	vertical cylindrical tanks allowed. at least 24" length from in to out. Length must exceed width if rectangular or oval tank. D = 30-72"	tee or baffle, at least 6" into liquid, but not below outlet tee	tee or baffle 40% into liquid for straight walls, 35% for other walls	inlet and outlet, all compartments must be accessible	steel, concrete, other materials			includes minimal installation and abandonment procedures. Require tank inspection every 6 years and pumping every 8 years.
Maine	extensive	750 (1-2 bd), 1000 (3-4 bd), 1250 (5 bd), 250 each adl. For others V = 1.5Q	multi compartment or tank required for institutional and commmercial over 2000 gpd, with first compartment at least 66%V. more than 2 compartments allowed for tanks over 1250 gal	d = 30" min. l = 74" min. 52" min diameter for vertical cylindrical	baffle or tee extending 20-30% into depth	baffle or tee extending at least 16" into liquid. Or effluent filter	to each compartment and above inlet/outlet. At-grade required for non-single family, encouraged for single family	precast/pour in place concrete, fiberglass or PE. Metal prohibited. Reference standards for each.	external grease int. required for food svc. Uses EPA formula. 750 gal min size. Design same as septic tanks, except outlet tee within 12" of tank bottom or effluent filter.		includes abandonment reg. and installation provisions. Septic tank degreasers prohibited. Additional measures if garbage disposals used. Appendix B for approval of proprietary devices including effluent filters.
Maryland	minimal	750 min for 2 bd, 250 for each adl., EPA formular for larger/non-res flows (roughly V = 1.5Q)							required for food svc facilities		no other specific design criteria in rule. May be part of MD policy.
Massachusetts	well-developed	V = 2Q, 1500 min	required for non-residential. In-series allowed. Two compartments max. first compartment V = 2Q, second V = Q	d = 4' min. rectangular tanks min 1.5:1 L:W ratio. Round tanks allowed, but min 6' from inlet to outlet. W = 3' min. vertical cylindrical tanks min diameter of 5'. Horiz cylindrical tanks l = 6' min and w = 3' min at water level.	tee at least 10" below liquid level	tee extending into liquid depth per table provided. gas baffle	w/in 6" of grade.	poured in place and precast concrete, fiberglass, PE. Metal not allowed.	required for food svc facilities. Min 24 hour HRT, min 1000 gal, min 4' depth. Inlet tee to mid depth, outlet tee to 12" off tank bottom. At-grade access.		manufacturers must have QA program in accordance with ASTM C1227. requirements for systems pumping to septic tanks. Some installation specs. Also, maintenance and abandonment regs.
Michigan	no state regs										Michigan has no state-wide regulations.
Minnesota	partially-developed	750 for 2 bd, 1000 for 3-4, 1500 for 5-6, 2000 for 7-9. increase by 50% and multi-compartment required for garbage disposals. Tanks over 3000 must be multi-compt, with V = 1.5Q for less than 1500 gpd, 1,125 + 0.75Q for greater than 1500 gpd	compartmentation allowed. For two compt tanks, first is 1/2-2/3 V. For three compt, first is 1/2, second and third are 1/4 each. In-series allowed with first tank larger and no tanks less than 1/4V. D = 78" max.	d = 30" min, all inside dimensions 24" min, l = 4' min.	tee or baffle, extending not more than 20% into liquid	tee or baffle, extending 40% (35% for cylindrical). Extend 50% into liquid for laundromats.	maintenance holes over all compartments w/in 6" of grade and access ports at grade	not allowed to be constructed of block, brick or similar materials that do not make a watertight tank			includes requirements for septic/dosing tanks and for pump to septic tank.
Mississippi	none										nothing could be found in MS regs for septic tank
Missouri	none										nothing could be found in MO regs for septic tank

State	Development of Regs.	Sizing	Compartmentation	Dimensions	Inlet	Outlet	Access	Construction	Grease traps	Preapproval	Other/Comments
Montana	well-developed	1000 (1-3b), 1500 (4-5b), 2000 (6-7b). V = 1.5Q for others up to 1500 gpd, 1125 + 0.75Q for above 1500 gpd		d = 78" max for < 5000 gal tank. For > 5000 gal tank, max d = 2.5L. L:W = 2:1. these don't apply to dose tanks	inlet tee or baffle extending at least 10% into liquid	effluent filter required. Extending 30-40% into liquid. Should be in baffle wall, if a two-compt. Combo septic/pump tank.	above inlet and outlet w/in 12" and marked with rebar.	concrete specs. All non-res must be watertested on site. Water, vacuum and pressure tests specified.	required for food svc.		includes section on effluent fiters. Requires NSF Std. 46 or documentation showing TSS removal of 30% min
Nebraska	extensive	1000 up to 3b, then 250 for each adl up to 10. for multiple dwellings on single tank, add one bd for each dwelling. Other facilities, use V = 1.5 Q up to 1500 and 1125+0.75Q for over 1500 . Double capacity for high strength wws.	two-compt required for > 2000 gpd, w/ 1/2-2/3 V in first. For over 2 compartments, first must be 1/2V and rest must be equally split. No more than four tanks in-series allowed, with first no smaller than others.	4' length min. d = 42-78" dia = 60" min. L:W = 2-3; 24" min interior dimension	tee or baffle 6" - 20% into liquid	outlet and baffle tee/baffle 40% into liquid (35% for cylindrical/horiz tanks.	manhole 6-12" below grade. Inspection ports above inlet and outlet with at-grade access.	concrete, concrete block, fibergass, HD plastic, FRP	required for food svc. Inlet extends 24" into liquid and outlet tee extends within 8" of bottom. 750 gal min, 24 hour HRT.		has detailed tank pumping reg and an abandonment reg. installatoin spec.
Nevada	partially-developed	1000 for 4 bd, 250 for each adl.	if used, first compt = 1/2-2/3V with outlet complying with outlet reqt.	2' min for all interior plan dimensions. D = 30-60"	tee or baffle at least 6" below liquid	tee or baffle 40% into liquid depth. Allows effluent filters with NSF certification	all compartments and inlet and outlet, within 18" grade	concrete, coated metal, vitrified clay, concrete block, brick. Interior of concrete tanks must be coated. Regs on precasting, built in place, coated metal (references UL-70 and Dept of Commerce 177-62) and homeowner construction septic tanks.		yes	some installation specs.
New Hampshire	partially-developed	1000 for 2 bd, 250 for each adl up to 10. add 50% for garbage grinder. 2X when raw sewage being pumped to tank. Other tanks, use 1.5Q up to 1500 and 1125+0.75Q over 1500	allow 2 compt, with first at 2/3V. 2 tanks may be used in series with first tank 2/3V.	d = 5 feet max up to 3000 gal and 6' max over 3000 gal.	vented tee extending at least 6" below liquid level, but not greater than outlet depth	vented tee extending 40% into liquid depth	to each compartment within 6" of grade.		required for non-res food service. 36 hr HRT min, 1000 gal min. Outlet baffle extending to within 6" of tank bottom		annual inspection required with requirement to pump the tank when sludge + scum is 1/3 liquid depth
New Jersey	well-developed	250 per bd, min 1000 gal. for other facilities, V = 1.5Q up to 1500 and 1250+0.75Q for over 1500	tanks in series allowed, provided first tank is no small than subsequent ones. Mult compt required for non-res over 1000 gpd. First compt 2/3V. Tanks over 1250 gallons may have more than 2 compts. Opening between compartments same as outlet.	d = 36-72"; rect tanks have L:W of at least 2:1; l = 72" min, w = 36" min at liquid level. Upright cylindrical tanks min dia of 52".	baffle or tee extending 25-33% into liquid	baffle or tee extending 25-40% into liquid. Gas deflector or effluent filter required.	over both compartments, within 6" of grade and inspection ports to grade over inlet and outlet.	precast concrete specs. Require interior coating of concrete tanks. Metal prohibited. Poured in place, precast concrete, fiberglass, PE allowed	required for food service, sized per EPA equation. 750 gallon min. Inlet/outlet tees extending to within 12" of tank bottom. Other requirements same as septic		some installation specs., abandonment spec.

State	Development of Regs.	Sizing	Compartmentation	Dimensions	Inlet	Outlet	Access	Construction	Grease traps	Preapproval	Other/Comments
New Mexico	well-developed	750 (1 bd), 1000 (2-3), 1200 (4), 1500 (5-6), 2000 (7-9). Also based on plumbing fixture units. Non-res: V = 2.5Q under 1500 gpd, 1125+0.75Q over 1500	2 compt mn required, with inlet being 2/3 total and at least 500 gal. second compt is 1/3 total, at least 5' in length for tanks over 1500 gal. Baffle shall have tee drawing from halfway point of liquid depth.	l = 5' min, w = 3' min. d = 2.5-6'	tee or baffle extending at least 12" below the water level	tee or baffle extending at least 12" below the water level	manhole over inlet and outlet within one foot of surface	concrete, concrete block, fiberglass, plastic. Lots of specs. Metal/wood prohibited.		yes	mostly specs IAPMO for septic tank design. ASTM for material specs.
New York	partially-developed	1000 gal up to 3 bd, then 250 for each adl., also has minimum liquid surface areas for various # bdrms. Add 250 and surface area for garbage grinder.	recommended. Required for tanks 10' long or more. First compt = 60-75% V. connect by slot or elbows extending 1/3 the depth into the liquid.	d = 30-60". Min 6' from inlet to outlet. L = 2-4 X width.	tee or baffle extending at least 12" below the water level when less than 40" d; 16" for d greater than 40"	tee or baffle extending at least 14" below the water level when less than 40" d; 18" for d greater than 40". Gas deflection baffle recommended, required when a garbage grinder.	within 12" of grade.	concrete, fiberglass, PE, steel meeting UL 70			
North Carolina	extensive	900 min for up to 3 bd, 1000 for 4, 1250 for 5, then V = 2Q up to 600 gpd, V = 1.17Q+500 for Q = 600-1500, V = 0.75Q + 1125 between 1500 and 4500 gpd and V = Q over 4500 gpd.	two compartment, with first being 2/3-3/4 total capacity. 2 tanks in-series allowed, provided that first is constructed w/o baffle wall and has at least 2/3 total capacity. Slot in baffle wall.	d = 36" min, L:W = 2 or greater	straight pipe, three blockouts, requires resilient connector	requires approved effluent filter with tee extending 25-40% into liquid depth. Requires resilient connector	over each compartment w/in 6" of grade.	extensive for precast concrete. Also cast in place and concrete block/brick masonry.	required for food svc. Sizing of 5 gal/meal , 2/3 required septic tank capacity or by EPA formula. Requires 2:1 L:W, two chambers, and baffle opening and outlet tee to extend ata least 50% into liquid. Access at-grade. Use of indoor interceptor can allow grease trap to be reduced by 50%.	yes	
North Dakota	partially-developed	1000 gal up to 3 bd, 1200 for 4 bd, 1500 for 5-6, 2000 for 7-8, 150+ for each adl bd. Also based on multiple dwelling units or apartments and number of fixture units.	up to four tanks in series allowed, with first tank at least as large as the others.	d = 30" - 6.5'. 24" horizontal dimension min., 4' between inlet and outlet min.	baffle or tee extending 6" to 20% into liquid	baffle or tee extending 40% of liquid depth (35% for horiz cylindrical)	within 6-12" of grade. Four inch min inspection pipe over inlet and outlet				where tanks installed above frost line, precautions must be taken to prevent freezing
Ohio	minimal	1000 gal for 1-2 bd, 1500 for 3 bd, 2000 (in 2 compts or tanks) for 4-5 bd, 2500 (in 2 compts or tanks) for 6 or more bd		d = 4' min.	vented baffle extending at least 6" into liquid, but not more than outlet baffle	vented tee, ell or baffle at least 18" below liquid.	at least one 10" dia cover to grade.				
Oklahoma	partially-developed	1000 gal up to 4 bd, then 250 per. Other facilities V = 1.5Q, with 1000 gal min.	1 or 2 compts. Baffle opening 20-40% of liquid depth. For two compt tanks, first shall be 1/2-2/3 V.	d = 3-6.5'	baffle extending at least 6" below liquid	baffle extending 20-40% below liquid.	manhole over each compartment and inspection port over inlet and outlet.	concrete, fiberglass and plastic (meeting IAPMO or CSA). Also mentions poured-in-place			

State	Development of Regs.	Sizing	Compartmentation	Dimensions	Inlet	Outlet	Access	Construction	Grease traps	Preapproval	Other/Comments
Oregon	well-developed	V = 2Q min for commercial; at least 1000 gal. For single family, 1000 gal min for up to 4 bd, 1500 for more than 4 bd.	two-compt required when tank preceded by sewage ejector pump. First compt at least 2/3V and effluent filter required. Tee in baffle with invert at same elevation as outlet tee.	d = 30-72", deeper for tanks over 3000 gal. distance between inlet and outlet shall be equal or greater than width.	tee extending at least 12" below normal high and low liquid levels. Gasket meeting ASTM c564	effluent filter with service riser. Opening extending 35-25% of lowest operating depth. Gasket meeting ASTM c564	at least one to grade with 20" diameter. 30" dia required for deep tanks. All tanks constructed to accommodate watertight risers. Sludge/scum measurement acceess at grade.	concrete, fiberglass or other non-corrosive materials approved by dept. cast-in-place allowed. Steel allowed per UL-70		yes	some installation specs. All tanks tested for watertightness using water test at site after installation. Require manufacturer to provide installation manual. Includes reg for dosing septic tanks.
Pennsylvania	partially-developed	900 gal min. table for various flows, culminating in 1.5Q for flows over 10,000 gpd	required with first compt or tank at least as large as second, but no more than 2X volume of second. Four tanks in series, max.	d = 3-7'; min horizontal dimension of 36"	baffle or vented tee at least 6" and no more than outlet tee in depth below liquid	baffle or vented tee of outlet of each tank or compartment shall extend 40% below liquid (35% for horiz cylindrical tanks)	over each compartment, within 12" of grade. 4" diameter inspection port above inlet tee brought to grade.	precast concrete, steel must meet US Dept of Commerce Standards 177-62. tanks may be constructed onsite if over 5000 gal, meet standards of Nat. Concrete Masonry Association (1957)			
Rhode Island	partially-developed	1000 up to 3 bd, add 250 for each adl bd. For other than individ dwellings, use 1000 up to 500 gpd, then V = 2Q up to 1500 gpd, then 1500 gal + Q above 1500 gpd	allowed, including 2 tanks in series, for capacity over 5000 gal provided first compartment is at least 1/2 total volume required	L:D = 1, min; L:W = 1.5, min. circular tanks at least 52" dia. D = 4-8'	tee or baffle extending at least 12" below water but no deeper than outlet	tee or baffle extending 1/3 of depth. Multiple outlets required for tanks with a width greater than 7'.	at-grade over inlet and outlet and each compartment	poured in place concrete, precast concrete, coated steel in accordance with Commerce 177 standard.	recommended for food svc facilities. Min d = 4', 1000 gal, at least 24 hr HRT. Inlet tee extend to mid-depth, outlet tee to within 12" of bottom. Baffles allowed. Access at grade		septic tank additives prohibited. Cannot pump into tank unless specifically approved. Some installation specs.
South Carolina	minimal	890 min. for 2 bd or less. 1000 for 3-4 bd, then add 250 for each adl bd. For other facilities, with flow of 593 or less, 890 gal, 593-1500 V = 1.5Q, 1500 or greater 1125+0.75Q							required for comm'l food prep establishments. Design in accordance with standards established by Health Authority.	yes	lint trap required on laundry sewer line preceding septic tank for com'l laundromats. Vehicle wash facilities required to have oil/water seperator on line preceding septic tank. Design per Health Authority.
South Dakota	partially-developed	1000 gal minimum up to 3 bd, then 250 for each adl bd. Add 20% for garbage disposal. For other systems with Q = 750-1500 gpd, V = 1.5Q; for Q >1500 gpd, V = 1125 + 0.75Q	two or more required for tanks larger than 3000 gal, with 1/2 to 2/3 capacity in first compartment; opening at same eleevation as effluent pipe and tees having same characteristics. Tanks in series allowed - first tank must not be smaller than subsequent tanks.	4' minimum horizontal dimension. D = 30-72".	baffle or tee extending 6" to 20% into liquid; not lower than outlet opening	baffle or tee extending 12" to 35% in cylindrical or 40% in rectangular tanks	over each compartment brought 6-12" of grade. Inspection pipes over inlet and outlet brought to grade.	coated metal tanks must meet SD plumbing code and cannot be used in systems when usage will be longer than 7 years. Minimal specs for concrete, fiberglass and plastic.	required for high oil/grease. Min size shall be 750 gal.		includes some installation procedures.

State	Development of Regs.	Sizing	Compartmentation	Dimensions	Inlet	Outlet	Access	Construction	Grease traps	Preapproval	Other/Comments
Tennessee	partially-developed	750 min for 2 bd, 900 for 3 bd, 1000 for 4 bd, add 250 for each adl bd. For others, up to 500 gpd, at least 750 gal; between 500-1500 gpd, V = 1.5Q; for > 1500 gpd, V = 1125 + 0.75Q	2 compartment with first 2/3-3/4 total. Opening same as outlet tee.	d = 30-60" under 3000 gal; up to 78" over 3000 gal.	three inlet blockouts. Inlet tee or baffle extending at least 12" below liquid level	one effluent blockout; tee extending 18" or 1/3 liquid depth whichever is less. Effluent filters may be used.	access over each compartment	rules written for precast. Others approved by department. Cast-in-place mentioned.	kitchen waste may be discharged to a grease trap. Properly submerged inlet and outlet. Sizing not specified.		
Texas	partially-developed		two or three compartments using baffles or multiple tanks. 1/2-2/3 in first compartment, opening between 25-50% of liquid level, using slot, hole or tee. For series tanks, first shall be at least 1/2 required volume (for two tanks) and 1/3 (for three tanks). first tank in a four or more tank systems shall be no less than 500 gal and the last tank shall contain no more than 1/3 total required volume.	d = 30 " min. Others in figure.	tee or baffle	tee. Effluent filters must meet NSF 46.	inspection or cleanout ports over inlet and outlet, 6-12" below grade.	reinforced cast in place, reinforced precast, fiberglass, polyethylene or other product approved by director. Metal prohibited. Concrete shall meet ASTM C1227	required for kitchen waste from institutions. Sized and designed in accordance with UPC 2000 and other standards approved by director		includes installation spec
Utah	extensive	750 min for 1 bd, 1000 for 2-3 bd, 250 for each adl. For other, up to 500 gpd, use 750 gal, for 500-1500 gpd, V = 1.5Q, for 1500-5000 gpd V = 1125 + 0.75Q	single or multiple. When compartmented, first must be at least 2/3 total required. No compartment less than 24" dimension. Baffle may have a slot 40% below liquid depth. Three compartments is max. For tanks in series, no tanks shall be smaller than 750. first tanks shall be at least 2/3 total required. no more than three in series	oval, circular, rectangular or square provided length at least equals depth. L:W should be at least 2-3. d = 30-72"	baffle or tee extending at least 6" deep, no deeper than outlet tee	may include effluent filter, at grade access required. Outlet baffle or tee extending to 40% (35% for horizontal cylindrical tanks). Multiple outlets prohibited. Gas deflector may be added.	access to each compartment within 4" of grade.	precast concrete, fiberglass, polyethylene, poured in place concrete, other approved by division. Has specs for precast concrete, fiberglass and PE and poured in place.		yes	requires watertightness testing per C1227 for all systems after installation. Some other installation provisions. Good section on maintenance, including advice about additives, waste brine, etc.
Vermont	partially-developed	1500 min for Q<750 gpd, for Q>750 gpd, V = 2Q. Larger if internal pump vault. 25% more if garbage grinder.			vented tee or baffle at least 6" below liquid level but not deeper than outlet	tee extending 40% (35% for horizontal cylindrical)	access over each compartment, at least one to grade	steel, concrete, fiberglass	required for food svc facilities. Designed for O&G < 25 mg/l. Interior interceptor does not substitute.		all septic tanks tested for leakage. Maintenance recommendations included also.
Virginia	well-developed	750 min for up to 2 bd, 900 for 3, 1200 for 4, 1500 for 5	can use 2-compartment instead of using effluent filter or providing at grade inspection port. Tanks in series allowed, provided first is 1/2-2/3 required volume.	rectangular in plan, cross-section and longitudinal. L:D:W should be at least 2:1:1 and less than 3:1:1. d = 4-8'	tee ofr baffle extending 6-8 inches below liquid	tee or baffle extending 35-40% into liquid	over inlet and outlet within 18" of grade				includes some installation. Maintenance spec requires at-grade inspection port or effluent filter or 30% larger tanks with two compartments with first compartment sized as indicated w/out 30% additional.

State	Development of Regs.	Sizing	Compartmentation	Dimensions	Inlet	Outlet	Access	Construction	Grease traps	Preapproval	Other/Comments
Washington	minimal	900 up to 3 bd, 1000 for 4 bd, 250 for each additional. For other V = 1.5Q with 1000 gal min					cleanout and inspection access within 12 inches of grade			yes	department maintains standards and lists of approved units.
West Virginia	none										nothing
Wisconsin	none										nothing
Wyoming	minimal	1000 gal up to 4 bd, then 250 for each adl. For other V = 1.5Q with 1000 gal min	can be partitioned, with first compartment at least 1/2 V.	L:W at least 2. d = 4-6'	tee or baffle	tee or baffle extending into middle 1/3 of water depth	access to each compartment and inlet and outlet. Cleanout over each compartment to grade.				

Most primary unit regulations appeared to share a common basis (US PHS and EPA Design Manuals) but there was still great variety in their level of development and specifics. Less than half (18) of the states had what would be considered "well-developed" primary unit regulations, although it should be noted that this doesn't necessarily mean state has a poor septic tank program.

As indicated previously, state regulations are generally slow to change. For example, most states still do not require multiple compartment tanks, though many encourage them and/or address them in regulations. Additionally, at least seven states allow vertical cylindrical tanks; the remainder address rectangular or horizontal cylindrical only. Seven states require effluent screens on all new tanks, while 2 additional states allow the use of an effluent screen to preclude a requirement to use some other septic tank enhancement. States addressing effluent screens typically reference NSF Standard 46 and/or maintain a list of state-approved products.

Grease traps are addressed in 21 states' onsite regulations and the grease trap is typically laid out similar to the septic tank design for that state, except with longer influent and effluent tees or baffles. Grease trap sizing is usually based on EPA or UPC sizing formulas or on a straight hydraulic residence time basis (usually 24 – 48 hours).

Thirteen state regulations specifically indicate that prefabricated tanks must be approved by the state. Several states had what appeared to be very well-developed programs. For example, in Alabama tank yards inspected twice per year, by regulation. The state of Massachusetts requires each manufacturer to implement a QA program in accordance with ASTM C1227. Florida also defines its tank QA program in its regulations. Other states may have good QA programs no described in their rules. For example, the QA program for North Carolina has been reported in the literature (D'Amato and Devkota, 1998), but is not explicitly codified in their state regulations. Additionally, many local regulatory jurisdictions undoubtedly have tank QA programs. These were not investigated.

Most regulations specify that tanks must be watertight, but few contain quality assurance measures to that end. Only three states' regulations require watertightness testing for all installed tanks, while one state requires testing in certain circumstances and two additional states describe watertightness testing, but do not appear to require it.

Development of Regulations

With regards to the overall development of each state's regulations, five qualitative ratings were used, as follows (the number of states' regulations falling into each category is indicated parenthetically):

- None (6) – nothing in the regulation addressing septic tanks or grease traps (this includes the two states with no state onsite regulations)
- Minimal (7) – includes something on septic tanks, but generally not detailed information
- Partially-developed (19) – these include a fair amount of design criteria on septic tanks, but may be lacking in one or more specific areas (construction, for example)
- Well-developed (11) – these generally include a complete suite of criteria on septic tanks (and often on grease traps also)

- Extensive (7) – these include a complete suite of criteria along with something that sets the program apart (at least by looking at the regulations); perhaps an exceptionally modern set of regulations or a well-developed QA program

Sizing

Most regulations contained basic sizing criteria for residential dwellings based on number of bedrooms. There was typically a minimum septic tank capacity that applied up to a certain number of bedrooms, with increasing sizes for more bedrooms. Above the minimum, an adder of 250 gallons per bedroom was common. At least one state also had a minimum surface area requirement in addition to minimum volumetric capacity and several states partially based septic tank sizing on number of plumbing fixture units. Most states included some criteria for non-dwelling or commercial/institutional septic tank sizing. Often, sizing was simply based on a straight multiplier of the design flow (e.g., a 24-48 hour hydraulic retention time), though many states utilized formulas for sizing septic tanks in different size ranges. Common sizing equations are listed below (where V is the required minimum septic tank capacity and Q is the design daily flow rate):

- $V = 1.5Q$ or $V = 2Q$, for Q up to a certain flow, usually 600-1500 gpd
- $V = 0.75Q + 1125$ for Q above 1500 gpd
- $V = Q$ for Q over a certain flow, e.g., 4500 gpd

Several states required larger capacities for dwellings with garbage grinders, high flow fixtures such as large tubs, and for pump-fed septic tanks. Several states had different sizing criteria for subsurface systems in different soil classifications or for facilities with strong wastes. A breakdown of the number of states with these stipulations in their septic tank sizing regulations is provided below:

- Garbage grinder sizing surcharge: 9
- High flow fixture sizing surcharge: 2
- Strong wastewater sizing criteria: 1
- Pump-fed septic tank sizing surcharge: 2
- Different criteria based on soil type: 1

Compartmentation

Requirements for septic tank compartmentation were very variable from state-to-state. Most states do not appear to require mandatory multiple-compartment tanks, though many encourage them and have regulations (dimensioning criteria) addressing multiple compartment tanks. Twelve states require multiple compartments for all septic tanks, while an additional eight require multiple compartment tanks in specific situations; usually above a certain capacity or for non-residential systems. Several states give an option of using a multiple compartment tank in exchange for not using an effluent screen, gas deflector or the like.

Where compartmentation is required or allowed, the sizing of the compartments is quite variable. Typical standards for the first compartment are 1/2, 2/3, or 3/4 of the total required volume (or some range encompassing these ratios). Multiple tanks-in-series are usually also allowed. Some states only allow a maximum of two tanks-in-series while others allow up to four, and some do not specify a maximum number. The opening between compartments is typically shown as a tee, although some states allow a slot or holes in the wall. The depth of the tee in the baffle walls of multiple compartment tanks, if required, is usually specified as being the same extension depth as for the effluent tee.

Dimensions

Dimensioning criteria is also very variable from state to state. Most states provide at least a minimum required depth, and a subset of those also include a maximum allowable depth (usually 6-8 ft), where liquid depths greater than the specified maximum do not count toward the required volume. A standard minimum depth is 36 inches. Also commonly specified are minimum length-to-width ratios and less commonly, length-to-depth ratios. Length-to-width ratio is commonly specified as being at least 2-to-1, although the specifications listed in the regulations reviewed ranged between 1.5-to-1 and 4-to-1. At least seven states still allow vertical cylindrical tanks where the length-to-width ratio is always 1-to-1. For states allowing vertical cylindrical tanks, a minimum diameter is typically specified ranging from 2 feet to 5 feet. Where length-to-depth ratios are specified (two states) they range from 1-to-1 to 3-to-1. A number of states specify minimum tank length (distance between inlet and outlet) and/or minimum interior horizontal dimension (usually 2 or 3 feet).

Most states specify a minimum freeboard/air space requirement of usually 15-20 percent of the total volume (sometimes expressed in inches of freeboard, rather than percentage of total tank volume).

Inlet and Outlet

Regulations regarding inlets and outlets are fairly consistent between state regulations. The vast majority of the states require a vented (open) tee or baffle extending above and below the liquid level by varying amounts (usually no deeper than the outlet tee). Only one state specifies a straight inlet pipe. We note that relatively few states distinguish clearly between the use of tees or baffles and, in some regulations, the words appear interchangeable.

Likewise, every state whose regulations cover this level of detail requires an outlet tee or baffle extending above and below the liquid level by varying amounts. A number of states specify that the outlet should extend 40 percent into the liquid depth for straight-walled tanks and 35 percent for cylindrical tanks. Many states allow a range of extension depths, expressed either in inches, percentage of liquid depth, or both.

Seven states require effluent screens on all new tanks, while two additional states allow the use of an effluent screen to preclude a requirement to use some other septic tank enhancement. Several other states address effluent screens in their regulations, but do not mandate their use. Most states addressing effluent screens reference NSF Standard 46 and/or maintain a list of state-approved products.

A number of states include criteria for the maximum (and less commonly, minimum) distance between the wall and the inlet or outlet tee or baffle. Most states specify a minimum fall between inlet and outlet inverts (usually 2-4 in).

Access

Access requirements vary greatly. Most states explicitly require access to all compartments and over the inlet and outlet, but the type, size and location with respect to finished grade vary considerably. There appears to be some conflict between providing access for convenient maintenance, while protecting against unauthorized entry, especially by children.

Construction

Most states include some construction specifications in their regulations. The majority most completely address precast reinforced concrete tank construction. Some states allow unreinforced concrete to be used if the wall thickness is a minimum of six inches. Built-in-place concrete or masonry block tanks are also frequently addressed, as are fiberglass and polyethylene. Where fiberglass and polyethylene tanks are addressed, the CSA and IAPMO septic tank standards are frequently referenced. Steel tanks are explicitly prohibited in five states and explicitly allowed in eight. Where allowed, two standards are commonly referenced: UL-70 and the U.S. Department of Commerce Standard 177-62 (note that this standard could not be located).

A number of states do not allow built-in-place tanks to be used in high groundwater conditions.

Anti-flotation provisions are cursorily addressed in a number of states' rules.

Grease Traps

Grease traps are addressed in 21 state onsite regulations. Where addressed, the grease trap is specified to be plumbed to receive only kitchen waste and grease trap effluent is to be plumbed into the septic tank inlet. Many of these states indicate that garbage grinders should not be plumbed to the grease trap.

Where grease trap design is addressed, the grease trap is typically laid out similar to the septic tank design for that state, except with longer influent and effluent tees or baffles; the effluent tee is often extended to within 12 inches of the tank bottom or closer. Sizing is usually based on the EPA or UPC formulas or on a straight hydraulic residence time basis (usually 24 to 48 hours).

Preapproval and Quality Assurance (QA) Programs

Thirteen state regulations specifically indicate that prefabricated tanks must be approved by the state. Other states may have approval or QA programs that are not codified in their regulations. Of the preapproval programs identified in the state rules, several states had what appeared to be very well-developed programs, most notably Florida and Alabama, which indicated that tank yards would be inspected at least two times per year. Massachusetts, even though its regulations do not indicate a prefabricated tank approval program, requires each manufacturer to have a QA program in accordance with ASTM C1227 and to seal their tanks as such.

Watertightness Testing

Most septic tank regulations specified that tanks must be watertight, but few contained any quality assurance to that end. Three states require watertightness testing for all installed tanks, one state requires testing in certain circumstances, and two additional states describe watertightness testing, but do not require it.

Other Issues

Most states include venting requirements; that is, an open/vented inlet is required with sufficient space in interior walls to allow gasses to pass freely through compartments and out the inlet pipe to be vented through the facility plumbing.

A number of states include installation procedures in their regulations and/or a simple abandonment specification. One state specified that tanks installed above the frost line must include precautions to protect against freezing, although those precautions were not listed.

Several states include maintenance requirements. Two states prohibit septic tank additives, while a few others mention them.

Two or three states specifically addressed design and sizing requirements for dosing septic tanks (septic tanks containing an effluent pumping system).

Texts and Guidance Manuals

A total of 26 popular texts and guidance manuals (listed in Table A-3) were reviewed, again in an effort to establish the current landscape of practice and to determine the most influential studies and references (via cross-referencing). Not surprisingly, the materials reviewed shared a lot of common information. In fact, it appeared that quite a lot of information was recycled from document to document without necessarily referring back to the reference serving as the original source of a given piece of information. In general, the number of original references that appear to have informed current guidance is quite small and most fundamental design criteria traces back to the PHS studies of the 1940-1950s.

Specific design criteria for septic tanks, including such details as compartmentation, sizing, inlet and outlet details, outlet penetration depth and others generally do date back to the PHS studies. Treatment efficacy guidelines are primarily based on data from several University of Wisconsin Small-Scale Wastewater Management Program (SSWMP) studies.

Onsite Wastewater Treatment Systems (Burks and Minnis)

This text contains an excellent history of septic tanks and septic systems, including a section on the history of septic tank additives and associated research, with comments about several states that have issued guidance on the use of additives.

Chapter 7 of the text is on treatment alternatives and includes a lengthy section on septic tanks. With respect to tank configuration, surface area is emphasized over depth, with wide shallow tanks preferred. Both one- and two-compartment septic tanks are described, with a statement that two-compartment tanks produce effluent with lower TSS and BOD. The text includes a section on grease traps and suggests that some jurisdictions allow the use of effluent screens in lieu of compartmentation.

Table A-3. List of Texts and Guidance Manuals Reviewed.

Title	Author	Publisher	Pub ID*
Septic Tank Practices	Warshall	Anchor/Doubleday	398
Septic Tank Systems: A Consultant's Toolkit	Winneberger	Butterworth Publishers	87
Decentralised Sanitation and Reuse	Lens, Lettinga, Zeeman	IWA Publishing	755
Treatment Wetlands	Knight, Kadlec	CRC Lewis	295
Advanced Onsite Wastewater Systems Technologies	Jantrania, Gross	CRC Press	10
Alternative Wastewater Collection Systems Manual	EPA/625/1-91/024	EPA	274
Onsite Wastewater Treatment and Disposal Systems	EPA/625/1-80/012	EPA	289
Onsite Wastewater Treatment Systems Manual	EPA/625/R-00/008	EPA	278
Response to Congress on Use of Decentralized Wastewater Treatment Systems	EPA/832/R-97/001b	EPA	270
Wastewater Treatment/Disposal for Small Communities	EPA/625/R-92/005	EPA	275
Septic Tank Systems: One Consultant's Toolkit	Winneberger	Hancor, Inc.	100
Onsite Wastewater Treatment Systems	Burks, Minnis	Hogarth House	753
Onsite Wastewater Disposal	Perkins	Lewis Publishers	649
Small and Decentralized Wastewater Management Systems	Crites, Tchobanoglous	McGraw Hill	271
Wastewater Engineering (Metcalf and Eddy), fourth edition	Tchobanoglous, Burton, Stensel	McGraw Hill	269
Wastewater Engineering (Metcalf and Eddy), third edition	Tchobanoglous, Burton	McGraw Hill	268
Consortium Educational Curriculum - Septic Tank Module	Seabloom, Bounds, Loudon	NDWRCDP	297
Model Decentralized Wastewater Practitioner Curriculum – Septic Tank Module	Loudon, Bounds, Converse, Konsler, Rock	NDWRCDP	802
1979 State of the Art Manual of On-Site Wastewater Management		NEHA	
The Septic Systems Owners Manual	Kahn, Allen, Jones	Shelter	272
Wastewater Engineering Design for Unsewered Areas	Laak	Technomic	71
International Source Book on Environmentally Sound Technologies for Wastewater and Stormwater Management	United Nations Environment Programme	UNEP-DTIE-IETC	290
Manual of Septic Tank Practice, 1958		U.S. PHS	631
Manual of Septic Tank Practice, 1967		U.S. PHS	632
Wastewater Treatment Plant Design	Vesilind	WEF/IWA Publishing	273
Considering the Alternatives	NC RCAP		276

* From associated bibliographic database

Chapter 9 of the text is on installation of onsite systems and includes good guidance on tank installation including provisions for excavation, bedding and backfill, as well as proper orientation and methods of making suitable pipe connections. Chapter 10 on operation and maintenance includes a small amount of information on septic tank maintenance.

Small and Decentralized Wastewater Management Systems (Crites and Tchobanoglous)

This comprehensive text is very commonly cited and contains an abundance of useful information on primary units in decentralized systems. Table 4-16 includes STE concentrations for septic tanks with and without effluent screens and with and without kitchen garbage grinders, with credit given to Bounds (personal communication, 1997).

Alternative sewer collection systems are discussed in Chapter 6, including a description of STEP systems using common septic/dosing tanks with screened pump vaults.

Chapter 5 includes an excellent discussion of gravity separation and settling theory, along with separate sections on septic tanks and grease traps. Septic tanks are described as a combined settling and skimming tank, an unheated unmixed anaerobic digester, and as a sludge storage tank. The accumulated sludge is described as being composed primarily of lint from clothes washing and lignous material contained in toilet paper. The text cites the more infrequent pumping of early septic tanks as evidence and suggests using lint traps and biodegradable toilet paper to limit sludge accumulation. The text suggests the use of single-compartment tanks fitted with effluent screens over two-compartment tanks. It also suggests that meander-type tanks with longitudinally placed baffles are a far more rational design than traditional width-wise baffles. Construction and watertightness testing provisions are introduced and sizing criteria for residential and larger flow tanks are provided. Sizing calculations for larger tanks are based in part on pump-out criteria developed by Bounds (1994).

Oil and grease removal is discussed in Section 5-11. Small grease interceptors are considered insufficient, while larger units – septic tanks modified with deeper inlet and outlet penetration points – have proven effective.

Section 5-12 discusses Imhoff tanks, which are essentially septic tanks where the liquid flow/solids settling and solids accumulation/digestion zones are physically separated via upper and lower compartments. The advantage is that solids resuspended by gas bubbles that evolve from digesting sludge do not contaminate the clarified liquid.

The Septic Systems Owners Manual (Kahn, Allen, Jones)

This manual is written with more of a lay audience in mind, but nevertheless contains some accurate, if not original, information as well as a host of visually appealing illustrations. The text features short narrative sections on a number of important and interesting issues like watertightness, building one's own septic tank (good idea?), considerations for cold climates, sizing/sludge storage, digestion, and inlet and effluent devices.

The authors seem to be influenced by Tim Winneberger's work, as they question the improved efficacy of two-compartment tanks and also promote the idea of the meander tank (despite there still – at this time of its publication – not being any actual data verifying performance). Credit to the idea of a septic tank with lengthwise baffles is given to Winneberger, while the name (meander tank) is credited to Warshall. The authors present anecdotal information that some engineers are designing and using meander tanks with success.

Chapters 5 and 6 address operation and maintenance and failure/troubleshooting, respectively. The information is well-researched and up-to-date (book was published in 2000) with reference to Bounds work indicating that pumping intervals of over 10 years may be more appropriate than the oft-repeated 3-5 years. Finally, the troubleshooting section recognizes structural failure of redwood tanks, belying the authors' California/Pacific Northwest affiliation.

Wastewater Engineering Design for Unsewered Areas (Laak)

Although it dates back to 1986, this decentralized wastewater management text, authored by University of Connecticut professor Rein Laak is still often referenced. Like the Crites and Tchobanoglous text, this is written as a formal engineering text and contains, in Chapter 4, excellent descriptions of septic tank function and design criteria, including a particularly nice presentation on BOD and TSS removal dynamics.

Figure 4-4 of the text shows two "efficient" septic tank designs, with one being two single compartment tanks-in-series and the other being a "University of Connecticut Trapezoidal Baffle Septic Tank" which is indicated to have 25 percent better removal of BOD and TSS than a conventional single-compartment tank. This unit appears to be a standard two-compartment tank fitted with a large lengthwise baffle set in the first compartment. The function of this unusual baffle is somewhat unclear and it is not described in the text.

Grease removal is also addressed with three types of passive treatment units illustrated, including a traditional grease trap (modified septic tank), and a European grease trap and gasoline oil trap. Chapter 6 is dedicated to septage management and includes septage quality characteristics.

Wastewater Engineering (Tchobanoglous, Burton, Stensel)

This is perhaps the most widely-referenced wastewater engineering text in use today. Undoubtedly owing to its authors' interests, the third edition contains good, general information about septic tanks; something of a synopsis of the Crites and Tchobanoglous content. While the fourth edition contains reasonable discussion on setting/sedimentation theory, it does not match Crites and Tchobanoglous in this area, nor does it contain any information specifically about septic tanks.

Onsite Wastewater Disposal (Perkins)

Another text specific to decentralized wastewater management, this 1989 volume includes a full section on septic tanks (in Chapter 4). Design criteria presented in this text directly references the Uniform Plumbing Code, although it doesn't appear that the UPC still addresses septic tanks. The Manual of Septic Tank Practice is also referenced for outlet penetration depth. Slotted baffle wall interconnections are suggested over pipe elbows to minimize clogging potential. General information on construction, installation and operation and maintenance is also presented.

Advanced Onsite Wastewater Systems Technologies (Jantrania, Gross)

This recent text makes the argument that septic tanks and drainfields alone are not sufficiently effective treatment systems and that advanced pretreatment is often necessary to adequately protect human health and the environment. Septic tank effluent characteristics are taken from EPA's Onsite Manual as well as the Crites and Tchobanoglous text. The authors argue that septic tanks have been extensively used without well-defined effluent quality standards. They also propose five treatment levels, with primary treatment – using existing data on septic tank effluent – constituting treatment level 1, and emphasize that effluent screens and routine maintenance are necessary to ensure that septic tanks consistently achieve these objectives.

Wastewater Treatment Plant Design (Vesilind)

This WEF/IWA wastewater design engineering text includes a full chapter (Chapter 5) on primary treatment, with accordingly good discussion of settling theory. Primary settling tank design considerations are discussed in some detail, including inlet feed and outlet withdrawal details and their impact on hydraulic performance.

Septic Tank Practices (Warshall)

Like the Kahn, Allen, Jones, book, Warshall's seems to be targeted to more of a field practitioner level audience rather than strict engineering. Chapter 5 specifically addresses septic tank design and includes sketches of one and two-compartment tanks, meander tanks and vertical cylindrical tanks-in-series (one of the various configurations tested by the PHS). As mentioned previously, Warshall coined the term meander tank, the design of which he claims that Winneberger "perfected" in the 1960s.

In a subsequent chapter on planning and construction, Warshall states that compartmented tanks are preferred with the three-compartment meander design best, followed by a three-compartment straight flow tank, followed by a two-compartment straight flow tank. No justification for the rankings are presented. Septic tank materials are also described, with redwood specified as lasting 30 years, the same as plastic and fiberglass, and more than precast concrete tanks, which are rated at a 20-year lifespan in the book. Tile cylinders (usually terracotta), metal and concrete block tanks are also described.

In a chapter on maintenance and care, the author suggests that the use of baking soda may buffer the septic tank environment and protect against wide pH fluctuation.

Consortium Educational Curriculum - Septic Tank Module

The development of these educational documents was sponsored by NDWRCDP and executed by the Consortium of Institutes for Decentralized Wastewater Treatment (CIDWT). The septic tank modules are extremely well researched, comprehensive and balanced guidance on septic tanks and the state of practice. There are two separate septic tank modules: one for an academic curriculum (297) and one for a practitioner curriculum (802). The documents are current, having been finalized in 2004 (academic) and 2005 (practitioner). Background on the documents is provided below in the words of CIDWT (http://www.onsiteconsortium.org/):

> "The CIDWT, often referred to as "The Onsite Consortium", is a group of educational institutions cooperating on decentralized wastewater training and research efforts. The Consortium also includes people from educational institutions, citizens groups, regulatory agencies and private industry.
>
> The Consortium of Institutes for Decentralized Wastewater Treatment has developed an educational curriculum as a joint effort between Consortium delegates from the academic and advisory communities including individuals from public agencies (state, counties and towns)

and private companies (consultants, manufacturers, engineers, designers soil scientists, and service providers).

The effort to develop a model curriculum was prompted by two educational projects involving over 40 training and education specialists in the on-site wastewater field from throughout the U.S. and Canada. The goal was to increase professionalism and improve the state of practice in the onsite/decentralized wastewater field. The projects are in the final development stages.

This work was supported [in part] by the National Decentralized Water Resources Capacity Development Project with funding provided by the U.S. Environmental Protection Agency through a Cooperative Agreement (EPA No. CR827881-01-0) with Washington University in St. Louis."

"The **Practitioner Curriculum** project is an effort to formalize and coordinate training opportunities for field practitioners in onsite/decentralized wastewater treatment. The goal of the curriculum is to:

- Improve field practitioners' abilities to effectively utilize onsite and decentralized technologies and improve the State of Practice (SOP).
- Enhance opportunities for the general public to make the best use of onsite/decentralized technology in their approach to a community's wastewater treatment needs.

This project provides a consistent technical educational base from which to launch nationwide training programs, but the modules available on CD may be modified to address specific local needs. The materials currently developed include a Model Practitioner Curriculum that can be used to organize training programs as well as four detailed modules.

The information posted online is largely in PDF format. For information on obtaining a CD of the materials, see the Introduction to the Practitioner Curriculum below.

The **septic tank module** (http://www.onsiteconsortium.org/training/activitiesDetails.cfm?ID=5) provides the materials needed to teach practitioners basic information on septic tanks. It is intended for use in courses offered with or without training centers. The target audience is onsite practitioners: soil scientists, system designers, installers, pumpers, regulators, maintenance personnel, etc. The information is presented in a stand-alone text document and illustrated with a PowerPoint presentation complete with instructor notes.

Typically, presentation of this module is preceded by a complete onsite system overview that defines and discusses all components of onsite systems. This provides the practitioner with perspective on the importance of septic tanks and how they fit into the treatment train."

"Under the **University Curriculum** project, appropriate modules were developed for teaching a one-semester laboratory and field course in onsite/decentralized wastewater treatment and natural water reclamation systems. The target audience for the materials is third- and fourth-year engineering students. The modules can also be adapted for undergraduate and graduate-level university courses in Environmental Health and other non-engineering curricula. The University Curriculum is available on CD-ROM with a navigational and organizational macro. The format is such that the materials are accessible and modifiable using software that instructors will have readily available.

The **septic tank module** (http://www.onsiteconsortium.org/files/seaseptic.htm) describes the history, materials of construction, and the physical, chemical, and biological processes that take place within the septic tank. It emphasizes that it is a marvelous cost efficient and energy free reactor and absolutely is a necessary part of small scale wastewater treatment systems. It accomplishes approximately fifty percent of the ultimate treatment without which downstream treatment schemes most certainly would fail.

The Course is intended for senior civil engineering and environmental health students who have previously completed an introductory course in environmental engineering.

The course goals/learning objectives are to make students aware that the much maligned septic tank is actually a marvel of simplicity and an energy free unit that provides the first and very important pretreatment of the wastewater in small scale wastewater treatment systems. To produce students who have a sound fundamental knowledge about the physical, chemical, and biological processes that take place within the septic tank. To inform them about the capability of the septic tank to renovate household wastewater and to inspire them to realize that it is the most important treatment unit in the small scale wastewater treatment system. The history, materials of construction, sizing, processes within the septic tank, sedimentation, theory, compartmentation, biological decomposition, aerobic decomposition, anaerobic decomposition, solid accumulation rate, septage pumping intervals, garbage grinders, buoyancy, effluent filters and screens, and additives."

Fact Sheets

A total of 57 relevant fact sheets and/or packages of fact sheets were collected and cursorily scanned for information pertaining to the performance of primary treatment units (mostly septic tanks) in decentralized systems in order to gain a better understanding of the current guidance being offered and the basis for that guidance, including references cited. While this does not constitute every such fact sheet in existence, the sample does give a good cross-section of the type and content of the information being communicated to practitioners and homeowners. A breakdown of the fact sheet sources is provided in Table A-4, along with the number of fact sheets reviewed from each source.

The fact sheets covered a number of related topics, including: general guides, septic tanks specifically, effluent screens, septic tank or system additives, tank or system maintenance (including care and pumping), and a few miscellaneous topics. Only a few fact sheets clearly indicated that their intended audience was the "homeowner", while many more did not explicitly state their audience, but appeared more broadly targeted toward diverse audiences that could include end users (homeowners), engineers and other professionals.

Table A-4. List of Fact Sheets Reviewed.

Source of Fact Sheet	Number Reviewed	Website
U.S. Environmental Protection Agency	7	http://cfpub.epa.gov/owm/septic/publications.cfm?program_id=70
North Carolina Cooperative Extension	1	http://www.ces.ncsu.edu/Publications/environment.php
National Onsite Water Recycling Association	1	http://www.nowra.org/?p=629
National Small Flows Clearinghouse	3	http://www.nesc.wvu.edu/nsfc/nsfc_products_archive.htm
Purdue University	4	http://www.ces.purdue.edu/henv/SepticSystems.htm
University of Arizona	4	http://cals.arizona.edu/pubs/
Montana State University	3	http://www.montana.edu/wwwpb/pubs/indexhomehealthfamily.html
University of Minnesota Extension	2	http://www.extension.umn.edu/OnsiteSewage/
National Precast Concrete Association	1	http://www.precast.org/
Ohio Department of Health	1	http://www.odh.ohio.gov/odhPrograms/eh/sewage/sewpubs/sewpubs.aspx
Texas A&M University	4	http://ossf.tamu.edu/septic_tank.html
Ohio State University	1	http://www.ag.ohio-state.edu/~agnrtech/waterpub.html
Dan Friedman	12	http://www.inspect-ny.com/septbook.htm
Wikipedia	1	http://en.wikipedia.org/wiki/Septic_tank
American Groundwater Trust	1	http://www.agwt.org/SepticSystems.htm
Sea Grant – Washington	1	http://www.wsg.washington.edu/research/ecohealth/septic.html
University of Rhode Island Extension	6	http://www.uri.edu/ce/wq/owtc/html/owtc_factsheets.htm
Northern Arizona University	1	http://www.cet.nau.edu/Projects/WDP/resources/SepticTanks.html
Penn State Cooperative Extension	3	http://www.abe.psu.edu/extension/factsheets/f/onlotsewageindex.htm

The researcher's cursory review sought to answer three major questions regarding each source:

1. How is a septic tank defined?
2. What goals, objectives, and/or purpose of the septic tank was stated?
3. What significant information about septic tank performance was included?

Definition of Septic Tank

For those fact sheets including a definition of the septic tank ("a septic tank is a…"), we expectedly found similarities between definitions. A compilation of the definitions results in a septic tank being:

- a tank, container, compartment, box, chamber or vessel;
- enclosed, buried, watertight;
- made of concrete, fiberglass or polyethylene (some also identified steel and redwood); and
- single or multi-compartment (or "series of tanks").

Some definitions also mentioned receiving sewage from the home and/or pretreating sewage prior to disposal in a drainfield.

A number of septic tank definitions included a range of volumetric capacities and/or typical size (length, width, height). When included, the sizes were clearly in the range of single-family capacity septic tanks, implying that the target audience is the homeowner and/or professional serving individual household systems.

Goals, Objectives and/or Purpose of Septic Tank

The stated goals, objectives and/or purpose of septic tanks as stated in the fact sheets reviewed were often extensions of the septic tank "definition". Some of the typical performance objectives were as follows (listed roughly in order of the more commonly mentioned to the less commonly mentioned):

- to settle out solids as sludge;
- to allow grease and oils and light solids to float to the surface as scum;
- to allow for biological digestion of solids (references range from 40 to 60 percent reduction);
- to vent gasses through plumbing piping/roof vents;
- to provide space for storage of sludge and scum;
- to allow flow of clarified effluent from the cleanest portion of the liquid column;
- to pretreat sewage and provide primary treatment of sewage; and
- to provide a consistent effluent that is easy to transport and treat.

Information on Performance

The information given on septic tank performance varies greatly according to, primarily, the specific topic of the fact sheet. In general, it appeared that fact sheets from extension services and universities and states were fairly state-specific. That is, the requirements of that particular state drove the recommendations or at least the manner in which the septic tank design criteria were presented.

Construction Materials

Most fact sheets only mentioned concrete, fiberglass and polyethylene tanks, but some referenced old steel, built-in-place concrete block and even redwood tanks, emphasizing their potential problems.

Several fact sheets emphasized the importance of watertightness.

Access

About half of the fact sheets strongly suggested that risers be used to provide at-grade access. Several fact sheets gave nice tips on how to locate a septic tank in the absence of at-grade access, but these did not necessarily recommend that at-grade access be provided after the tank was located.

Compartmentation

Many fact sheets showed or described either only one-compartment tanks or both one- and two-compartment tanks. Those that mentioned both one- and two-compartment tanks usually stated that two-compartment tanks were preferred. One source indicated that two-compartment tanks were more maintenance intensive and generally larger, although the basis for this contention is specious at best. Several showed only two-compartment tanks, indicating that they were required in their state. Few of the sheets indicated the relative sizing of the compartments, although most showed or stated that the first compartment should be one-half to two-thirds of the total capacity.

Influent/Effluent Structures

The majority of the fact sheets depicted septic tanks as having both inlet and outlet baffles or tees. Many strongly suggested inspecting the quality of the baffles, especially if constructed of metal or concrete. Several recommended replacing such baffles with PVC tees. Few fact sheets showed a septic tank with a straight inlet. About half of the fact sheets addressing the outlet structure strongly suggested that effluent screens be used; a few of the fact sheets did not mention effluent screens at all. An EPA fact sheet indicated that as of its printing, 50 counties in the U.S. and the states of Florida, Georgia, North Carolina and Connecticut required the use of effluent screens.

Additives

The fact sheets were quite consistent in their recommendations on additives, suggesting that they were almost uniformly unnecessary. Several fact sheets went into significantly more detail on additives, splitting them up into inorganic, organic solvents and biological additives. In general, they state that biological additives are unlikely to be harmful, but chemical additives may be. There were several warnings that phosphorus removal chemicals could reduce buffering capacity, causing pH to plummet and the tanks biology to die. Several references are cited to substantiate information regarding additives, including the Clark thesis (North Carolina State University) and other fact sheets, including the University of Rhode Island's (URI), which cross-references the Clark thesis as well as the University of Arkansas and West Virginia University studies.

Large Flow/Non-Residential Septic Tanks

Only one fact sheet we reviewed addressed non-residential or large flow systems in any detail and it included very little useful design information regarding the septic tank. Vague design statements that were made (e.g., suggesting the use of three septic tanks in series for restaurants) were unsubstantiated and unreferenced.

Dos and Don'ts

Most of the fact sheets, particularly those clearly targeted to homeowners/end users, included guidance on caring for the septic system and for maintenance. Several fact sheets indicated that the use of garbage grinders (disposals) can increase the solids loading to the tank by up to 50 percent. Overall, the dos and don'ts were generally consistent between fact sheets, with some being much more detailed and/or creative than others.

Shape

Most tanks in the fact sheets were depicted as being rectangular in plan view, although several indicated that circular tanks could be used, and at least one stated that size was more important than shape. Several fact sheets stressed the importance of aerial size (length x width) on settling efficacy, although they were less clear about the trade-off of a shallower depth (with respect to storage, etc.)

Sizing

Most fact sheets indicated that a hydraulic residence time of at least 24 hours was typical. Several presented sizing criteria as required by their state. Many simply stated that the sizing of a septic tank is based on the number of bedrooms (again implying that the primary audience is the homeowner or other individual focusing on residential systems). Several references made the worthy point that solids accumulation reduces the hydraulic residence time, potentially impacting performance.

Tank Pump-Out Criteria

Tank pumping frequency and/or physical criteria upon which to base pumping varied greatly among fact sheets. On one hand, many fact sheets reprinted a table listing frequency as a function of tank and household size. Those that referenced the table generally credited a Penn State Cooperative Extension publication or a table from a publication by Mancl (1984). The Penn State publication indicates that their table of estimated pumping frequency is based on maintaining a minimum 24 hour hydraulic retention time assuming 50 percent digestion of retained solids. Most fact sheets gave ballpark pump-out frequencies which varied greatly and were sometimes inconsistent with other criteria given in the fact sheet (like the aforementioned table). In general, the frequency ranged from annually to every 5-7 years. Most fact sheets appropriately emphasized the importance of routine inspection and sludge/scum depth measurement upon which to base pumping events.

Despite this, the physical criteria upon which to base pumping need also varied greatly. Some criteria were based on a percentage of the liquid depth occupied by the settled sludge, while others were based on the distance between the settled sludge blanket and the effluent tee entrance. When a percentage criteria was stated for the sludge depth, it was between 25 and 33 percent of the total liquid depth. When a distance from the effluent tee was stated, it was 12 inches. There was less agreement for scum distance from the effluent tee, with some guidance using 3 inches and some using 6 inches.

Cited References

Most fact sheets did not diligently cite references. Those that did often cited other fact sheets. Our objective, however, was to find the basis for the information presented in the fact sheets.

Clearly, as stated above, an implicit reference was usually the host state's regulations. Several fact sheets, however, were nationwide in scope.

The most common general references were the following:

- U.S. EPA Onsite Design Guidance Manual (2002)
- Crites/Tchobanoglous text
- Minnis/Burks text

References specific to certain topics (additives, pumping frequency, etc.) are indicated in the subsections above.

REFERENCES

Bounds, T.R. 1994. *Septic tank sizes for large flows*. Proceedings, Seventh International Symposium on Individual and Small Community Sewage Systems, ASAE, December 11th - 13th, 1994, Atlanta, GA.

D'Amato, V.A., I.C. Devkota, 1998. *Development of Prefabricated Septic and Pump Tank Construction and Installation Standards for North Carolina.* Proc. Eighth National Symposium on Individual and Small Community Sewage Systems, American Society of Agricultural Engineers, presented in Orlando, FL, March 1998.

Mancl, K.M. 1984. *Estimating Septic Tank pumping Frequency.* Journal of Environmental Engineering.

APPENDIX B

HISTORY/EVOLUTION OF PRIMARY TREATMENT UNITS IN DECENTRALIZED SYSTEMS

Several texts and guidance manuals provide excellent histories of sewage treatment and the evolution of decentralized systems in particular. Texts by Minnis and Burks (1994), Winneberger (1984), and Crites and Tchobanoglous (1998), the training module by Seabloom, Bounds and Loudon (2005), and the international text by Lens, Lettinga and Zeeman (2001) provide consistent and thorough accounts of the history of septic tanks which are widely accepted and reused. Minnis and Burks (1994) and Winneberger text (1984) provide the most comprehensive historical overview specific to the septic tank.

These texts describe the existence of drains to transport water away from buildings and to cesspools (the earliest predecessor of the septic tank) in India, Pakistan and the Island of Crete dating back to 2000 B.C. and earlier. In fact, Minnis and Burks (1994) quote the Old Testament, where Moses lays out perhaps the first recorded "code" of sanitation in the Book of Deuteronomy (23:12-14). Open Roman sewers transported untreated sewage to the River Tiber as early as the 6th Century B.C. and the Ancient Greeks started conveying liquid waste to agricultural fields several hundred years later.

Driven by epidemiological studies tying disease outbreaks to the contamination of wells by sewage in the mid-1800s, water began to be used in larger quantities to transport wastes away from the population and into surface waters. Furthermore, the introduction of the flush toilet by Thomas Crapper & Co. around 1872 and the accompanying volumetric increase in waste generation spurred the development of new technologies for handling wastewater. As such, several predecessors of the septic tank were developed in the mid-1800s in England and Germany. These were relatively large semi-centralized units. Lens, Lettinga and Zeeman (2001) credit the first primary settling tanks to the digging of settlement pits at Craigentinny Meadows in Edinburgh in 1829. Flat-bottomed tanks came next, some of which had clay-lined bottoms, but most references attribute the "invention" of the septic tank to Louis M. Mouras although he patented his unit as the Mouras' Automatic Scavenger. The invention was picked up by a French scientific journal, Cosmos les Mondes, in 1881 where its editor suggested that anaerobic microorganisms were responsible for liquefying the solids collected in the air-tight tank. The invention was later described in the U.S. by The Engineering News (April 15, 1882).

The term "septic tank" was coined by Donald Cameron who had designed a watertight covered basin to anaerobically treat sewage in Exeter, England in 1895. Cameron received a U.S. patent on his septic tank in 1899. When he attempted to collect royalties on his patent in the U.S., Leonard Metcalf challenged the patent by researching the history of septic tanks in a paper entitled, "Antecedents of the Septic Tank" (1901). Metcalf found numerous examples of variations of the design that had previously been employed in England, continental Europe and the U.S. The Minnis and Burks text includes copies of the original drawings of septic tank precursors from the Metcalf paper.

In 1908, Dr. Dunbar (no first name given) of the Hamburg (Germany) State Hygienic Institute published "Principles of Sewage Treatment" where, in addition to sections on the removal of suspended solids and organics from sewage, he described experiments that could be considered the first controlled research study on septic tank processes. The experiments would be considered unorthodox by today's standards (for example, he submerged dead animals in septic tanks and tracked their degradation over time), they did establish at least the idea that septic tank treatment is a biologically robust process capable of significant liquefaction of solids (the animals decomposed rather quickly). Dunbar also addresses the issue of grease and oil removal. He states introductorily, "…it is the custom in sewered towns to insist upon the provision of grease traps at all places where grease is likely to enter the sewers…" and proceeds to describe several innovative grease removal devices that can be used to enhance removal above that provided by a grease trap (not described). It is interesting to note that the devices illustrated in Dunbar's publication appear to be considerably more complex—from an engineering standpoint—than the grease traps and septic tanks in use today.

Pre-War (WWII) America saw relatively rapid industrialization and increases in population, both via immigration and native births. Pioneering and colonization of uninhabited areas continued. Centralized sewage collection and primary treatment systems were developed in the major cities, while homesteaders, farm families and others in less populated towns were left to manage their sewage onsite. Decentralized system technologies developed mostly out of necessity, rather than according to regulatory codes or even well-established guidance. Industrialization increased exponentially during the war effort of the late 1930s and 1940s and, after the dust had settled, the United States experienced a period of unprecedented growth—in population as well as economically, in Americans' quality of life. This prosperity resulted in a building boom which expanded the population into exurban (soon to be suburban) areas. In many cases, the building and home financing was subsidized by the federal government in the form of low-interest mortgages and the like. This combination of increased pressures (in terms of population and land use) and increased expectations (in terms of quality of life) resulted in decentralized wastewater systems being more actively considered. Nevertheless, treatment system components, now being mostly conventional septic tank/soil absorption systems, cesspools or straightpipes, were typically designed based on convenience and material availability.

Notwithstanding the efforts of the inventors of the septic tank antecedents of the late 19th century, and the efforts represented by several limited studies in the early 20th century, relatively little thought had gone into the engineering design of units serving decentralized systems until efforts put forth by PHS starting in the 1940s. The Joint Committee on Rural Sanitation (JCORS) was formed in 1941 by the PHS. In 1954, the group consisted of representatives of the PHS, USDA, the Conference of Municipal Public Health Engineers, the Conference of State Sanitary Engineers, the Federal Security Agency, the Housing and Home Finance Agency, Tennessee Valley Authority (TVA), U.S. Department of Interior, and the Veterans Housing Administration. They originally published their recommendations for septic systems in 1943, a revised version in 1947, and then an unrevised version with overall strategic suggestions was added in 1950 (Kreissl, personal communication).

Following WWII, the Federal Housing and Home Finance Agency (HHFA) recognized the exurban housing trends and the importance of decentralized sewage systems, frequently being saddled with houses on failing septic systems and sometimes on mortgage defaults due to families leaving such homes. In 1946, the HHFA initiated studies aimed at developing "a factual basis on which (onsite systems) could be designed, installed and maintained." HHFA jointly funded a number of seminal studies conducted by the PHS in the late 1940s and 1950s and, later, by the UC – Berkeley Sanitary Engineering Research Lab in the 1950s and 1960s. The PHS work included extensive laboratory/pilot experiments and field survey work focused on establishing scientifically-based design parameters for septic tanks. The PHS also conducted work on soil absorption systems. The UC-Berkeley work had a greater focus toward the soil absorption component of conventional onsite systems, but also included some significant work on septic tank design and operation. Later work was administered by the U.S. Environmental Protection Agency (U.S. EPA) and was conducted by several researchers in the 1970s, with the bulk of the work being done by the University of Wisconsin Small-Scale Wastewater Management Program (SSWMP). This work had several foci, although the further development of primary treatment unit technology was not a primary objective. Continued work on soil absorption systems, development of sand filtration technologies, graywater segregation, septage characterization (not done by SSWMP), the study of alternative collection systems (septic tank effluent pump (STEP) and grinder pump systems) and other topics were studied. Relevant publications generated from these research efforts are discussed in the subsections below as well as in the appropriate sections of subsequent chapters.

Through the decades of the 50s, 60s, and 70s, design criteria for decentralized system technologies, including septic tanks, solidified and was adopted into numerous guidance materials, including several revisions of the PHS "Manual of Septic Tank Practice", between 1957 and 1969. JCORS were advisors to the PHS when they issued the first Manual of Septic Tank Practice in 1957. The membership of JCORS had changed to include what is now the Water Environment Federation (WEF), the U.S. Coast Guard, and the American Public Health Association (APHA) (Kreissl, personal communication). In the 1970s, the SSWMP developed some guidance materials for EPA, culminating in the first version of EPA's onsite design guidance manual in 1980 ("Design Manual, Onsite Wastewater Treatment and Disposal Systems"). The Manual was revised most recently in 2002 as the "Onsite Wastewater Treatment Systems Manual". The adoption of standard design practices over the years resulted in codification into state regulations as well as industry standards and engineering specifications. With few exceptions, state regulations drive current practice.

Over the years, a host of less concentrated work has been done to both answer outstanding questions regarding septic tank design and to advance the state of primary treatment unit technology.

The primary treatment units of the 21st century have greater demands upon them than their predecessors. An ever-increasing array of household chemicals, personal care products and pharmaceuticals has the potential to stress their biological function and to contaminate receiving environments if insufficiently treated. Trending away from semisolid greases toward the use of liquid oils for cooking has major implications on separation processes in primary units. New water using appliances may have different hydraulic profiles, and therefore consequences, on primary unit performance. Likewise, the impact of point-of-use water treatment systems on decentralized wastewater systems continues to raise questions in the industry. As such, additional work on primary units continues today in various areas. The research is somewhat more focused on specific issues/questions compared to studies such as those conducted by the PHS in the mid-20th century, and the work is being done by a greater variety of researchers. Not surprisingly, a significant amount of today's work is funded by, and often conducted by, the private sector.

REFERENCES

Crites, R. and G. Tchobanoglous. 1998. *Small and decentralized wastewater management systems.* McGraw-Hill.

Dunbar, C.G. 1908. *Principles of sewage treatment.* Charles Griffen and Company, Ltd.

Lens, P., G. Lettinga and G. Zeeman. 2001. *Decentralised Sanitation and Reuse.* IWA Publishing.

Metcalf, L. 1901. *The Antecedents of the Septic Tank.* ASCE, Vol. XLVI, pp 456-490.

Minnis, M.M. and B.D. Burks. 1994. *Onsite Wastewater Treatment Systems.* Hogath House, Ltd.

Seabloom, R.W., T.R. Bounds and T.L. Loudon. 2005. *University curriculum development for decentralized wastewater management.* NDWRCDP Report.

Winneberger, J.H.T. 1984. *Septic-tank Systems A Consultant's Toolkit. Volume II – The Septic Tank.* Butterworth Publishers.

GLOSSARY

Anoxic reactor: A septic tank into which nitrified secondary effluent is returned to facilitate denitrification using the organic carbon and low dissolved oxygen levels in the tank.

Baffle: Physical barrier placed in a tank to dissipate energy, direct flow, and/or create compartments within a tank.

Biogas ebullition: The formation and release of gas bubbles from the settled sludge layer in a septic tank as a result of anaerobic digestion in the layer, which may resuspend settled solids, hinder settling, and/or seed the clear zone and scum layer with microorganisms.

Closed-conduit, laminar flow tank: A proprietary septic tank design by Waterloo Biofilter, based on the meander tank design and featuring no headspace.

Compartmentation: The division of a single septic tank into compartments using baffle walls.

Computational fluid dynamics: A branch of fluid mechanics that uses numerical methods and algorithms to analyze problems that involve fluid flows.

Cumulative frequency diagram: A graphical depiction of constituent concentrations in relation to cumulative frequencies of occurrence, expressed as a percent.

Decentralized wastewater system: Collection, treatment, and dispersal/reuse of wastewater from individual homes, clusters of homes, isolated communities, industries, or institutional facilities, at or near the point of waste generation.

Effluent screen: Easily removable, cleanable, or disposable device installed on the outlet piping of a septic tank or grease trap for the purpose of retaining solids larger than a specific size and/or modulating effluent flow rate. Also sometimes called an ***effluent filter***.

Gas deflector baffle: Baffle designed to direct gases and rising solids away from the bottom of the outlet.

Grease trap: Relatively large device similar to a septic tank located outside a facility that generates commercial food service wastewater and is designed to intercept, congeal, and retain or remove FOG.

Hydraulic efficiency: The ratio of actual hydraulic residence time (as measured using tracer studies) to the design HRT (volumetric capacity/flowrate).

Imhoff tank: Settling tank first designed by Karl Imhoff in 1906, similar to a two-compartment septic tank with compartments separated vertically, with flow through only the top, or settling compartment. Settled solids slide down inclined walls into a lower digestion compartment (typically with a sloped hopper-type bottom) and internal baffling prevents biogas-driven resuspension of solids into the clear settling zone

Intercompartment transfer device: The device through which liquor in the septic tank flows between adjacent compartments; usually a slot, tee, orifice, or similar.

Internal pump chamber: A pump typically housed within a screened chamber that is placed at the outlet of a septic tank for pumping effluent to downstream system components.

Meander tank: A septic tank with longitudinally-placed baffle walls rather than baffles placed perpendicular to the flow, as is typical practice.

Primary treatment: Physical treatment processes involving removal of particles, typically by settling and flotation; a grease interceptor and/or a septic tank typically provide primary treatment in decentralized systems.

Septic tank: A water-tight, covered receptacle for treatment of sewage; receives the discharge of sewage from a building, separates settleable and floating solids from the liquid, digests organic matter by anaerobic bacterial action, stores digested solids through a period of detention, allows clarified liquids to discharge for additional treatment and final dispersal, and attenuates flows.

Tanks-in-series: Orienting multiple tanks in a series flow arrangement to increase volumetric capacity or effectively provide compartmentation.

UASB-septic tank: Septic tank featuring a series of compartments separated by baffle walls where effluent from one compartment is directed downward to flow up through the settled sludge blanket of the next compartment in series. Also sometimes called ***baffled anaerobic reactors*** or ***baffled anaerobic septic tanks***.

Surface loading rate: Flowrate divided by the surface area of a settling chamber (septic tank).

Sludge seeding: The practice of adding sludge or septage to a septic tank to initiate biological activity or simulate the accumulation of settled sludge (e.g., during research studies).

REFERENCES

Anderson, D.L., A.L. Lewis, and K.M. Sherman. 1991. *Human Enterovirus Monitoring at Onsite Sewage Disposal Systems in Florida.* Proceedings of the Sixth National Symposium on Individual and Small Community Sewage Systems, Amer. Soc. Ag. Engineers, Chicago, IL, December 16 - 17, 1991.

Baumann, E.R. and H.E. Babbitt. 1953. *An Investigation of the Performance of Six Small Septic Tanks.* University of Illinois Press: Champaign, IL, 1953.

Baumann, E.R., E.E. Jones, W.M. Jakubowski, and M.C. Nottingham. 1997. *Septic Tanks.* Proceedings of the Second National Home Sewage Treatment Symposium, Amer. Soc. Ag. Engineers, Chicago, IL, 1977, Publication 5-77.

Bitton, G. 1999. *Wastewater Microbiology, Second Edition.* Wiley Series in Ecological and Applied Microbiology. Wiley & Sons: New York, 1999.

Bounds, T.R. 1982. *Wastewater Characteristics: Glide, Oregon, Pressure Sewer System.* Report of the Douglas County (OR) Public Works Department: Douglas County, OR, 1982.

Bounds, T.R. 1988. *Glide Audit 1986-1987 Summary of Sludge and Scum Accumulation Rates.* Report of the Douglas County (OR) Public Works Department: Douglas County, OR, 1988.

Bounds, T.R. 1993. *Alternative Sewer Design Workshop: Effluent Sewer Technology Septic Tank Effluent Pump (STEP) and Septic Tank Effluent Gravity (STEG)Ssystems.* Alternative Sewer Design Workshop, 1993.

Bounds, T.R. 1994. *Septic tank sizes for large flows.* Proceedings of the Seventh International Symposium on Individual and Small Community Sewage Systems, Amer. Soc. Ag. Engineers, Atlanta, GA, December 11 - 13, 1994.

Bounds, T.R. 1996. *Alternative sewer designs: effluent sewer technology, septic tank effluent pump (STEP) and septic tank effluent gravity (STEG) systems.* On-Site Wastewater Treatment, 5th Annual Conference, 1996.

Bounds, T.R. 1997(a). *Design and performance of septic tanks.* Site Characterization and Design of On-Site Septic Systems, ASTM STP 1324, M.S. Bedinger, J.S. Fleming, and A.I. Johanson, Eds., American Society for Testing and Materials.

Bounds, T.R. 1997(b). *Septic tank septage pumping intervals.* Texas Onsite Wastewater Treatment Research Council Conference, Plano, TX, April 13 - 15, 1997.

Bounds, T.R., G. Denn, and N.T. Bounds. 2004. *Maintaining and troubleshooting advanced onsite systems: git 'er done!* Proceedings of the National Onsite Water Recycling Association's 13th Annual Conference, Albuquerque, NM, November 7 - 11, 2004.

Boyer, J.A. 1992. *The Effects of Septic Tank Modifications on Effluent Characteristics.* University of Maine Press: Orono, ME, 1992.

Brandes, M. 1977. Effective Phosphorus Removal by Adding Alum to Septic Tank. *J. Water Pollut. Control Fed.* November 1977, *Vol. 49, No. 1, pp. 2285-2296.*

Brandes, M. 1978. Characteristics of Effluents from Gray and Black Water Septic Tanks. *J. Water Pollut. Control Fed., Vol. 50, No. 11, pp. 2547-2559.*

Brooks, B. 2006. Proceedings of the National Onsite Water Recycling Association, 15th Annual Conference, Denver, CO, August 28 - 30, 2006.

Brown, C.A., K.E. Kiernan, J.F. Ferguson, and M.M. Benjamin. 1984. *Treatability of Recreational Vehicle Wastewater in Septic Systems at Highway Rest Areas.* NTIS Report. Transportation Research Board, Transportation Research Record No. 995, ISSN: 0361-1981, Washington DC.

Burks, B.D. and M.M. Minnis. 1994. *Onsite Wastewater Treatment Systems.* Hogarth House, Ltd.: Madison, WI.

Byers, M.E., K.E. Zoeller, and J.D. Fletcher. 2001. *Septic Tank Effluent Filters: Monitoring and Assessment.* Proceedings of the Ninth National Symposium on Individual and Small Community Sewage Systems, Amer. Soc. Ag. Engineers, Fort Worth, TX, March 11 - 14, 2001.

Camp, T.R. 1946. *On Sedimentation and the Design of Settling Tanks.* Trans. ASCE, Vol. 111, pp. 895-936.

Carroll, J.A. 2005. *Lessons Learned from a Model 5 EPA Management Program for Onsite Wastewater Systems.* Proceedings of the National Onsite Water Recycling Association, 14th Annual Conference, Cleveland, OH, October 10 - 13, 2005.

Christopherson, S.H., D.G. Gustafson, and J. Anderson. 2006. *Preliminary Evaluation of Cluster System Septic Tank Performance.* Proceedings of the National Onsite Water Recycling Association, 15th Annual Conference, Denver, CO, August 28 - 30, 2006.

Clark, G.H. 1998. *The Effect of Bacterial Additives on Septic Tank Performance.* Master's Thesis, North Carolina State University, Raleigh, NC, 1998.

Cooper, I.A. and J.W. Rezek. 1978. *Treatability of Pressure Sewage.* Proceedings of the Fifth National Conference on Individual Onsite Wastewater Systems, NSF, Ann Arbor, MI, October 1978, pp. 121-154.

Crites, R. and G. Tchobanoglous. 1998. *Small and decentralized wastewater management systems.* McGraw-Hill: New York, 1998.

D'Amato, V.A. and I.C. Devkota. 1997. *Development of Prefabricated Septic and Pump Tank Construction and Installation Standards for North Carolina.* Proceedings of the Eighth National Symposium on Individual and Small Community Sewage Systems, Amer. Soc. Ag. Engineers, Orlando, FL, March 8 - 10, 1998.

DeJong, K.E., R.L. Siegrist, L.B. Barber, and A.L. Wren. 2004. *Occurrence of Emerging Organic Chemicals in Wastewater Effluents from Onsite Systems.* Proceedings of the Tenth National Symposium on Individual and Small Community Sewage Systems, Amer. Soc. Ag. Engineers, Sacramento, CA, March 21 - 24, 2004.

DeWalle, F.B., D.A. Kalman, D. Norman, J. Sung, and G.D. Plews. 1985. *Determination of Toxic Chemicals in Effluent from Household Septic Tanks.* US Environmental Protection Agency, Washington, D.C., EPA/600/2-85/050.

Dunbar, C.G. 1908. *Principles of Sewage Treatment.* London Press, 1908.

Dziegielewski, J.O., J.C. Kiefer, E.M. Opitz, G.A. Porter, G.L. Lantz, P.W. Mayer, W.B. DeOreo, and J.O. Nelson. 2000. *Commercial and Institutional End Uses of Water*. American Water Works Research Foundation, Denver, CO.

Edwards, D. and A. Decarvalho. 1998. Assessing the impact of household cleaning products on wastewater treatment systems. *The Small Flows Journal*, Vol. 4, Issue 1, pp. 15-27.

Elmitwalli, T.A., L. Groendjik, J. van Lier, G. Zeeman, and G. Lettinga. 2003. Decentralised treatment of concentrated sewage at low temperature in a two-step anaerobic system: two upflow-hybrid septic tanks. *Water Sci. Technol.*, Vol. 48, No. 6, pp. 219-226.

Engle, D. 2006. Coming Soon: NOWRA's Model Code. *Onsite Water Treatment*. Vol.1, No. 2. September/October 2005.

Etnier, C., D. Braun, A. Grenier, A. Macrellis, R.J. Miles, and T.C. White. 2005. *Micro-Scale Evaluation of Phosphorus Management: Alternative Wastewater Systems Evaluation*. Project No. WU-HT-03-22. Prepared for the National Decentralized Water Resources Capacity Development Project, Washington University, St. Louis, MO, by Stone Environmental, Inc., Montpelier, VT, 2005.

Florida Department of Health. 1996. *The Determination of Several Effluent Properties from Food Service Establishments that Employ On site Sewage Treatment Systems - Final Report Phase I.* State of Florida Dept. of Health, Bureau of Water and Onsite Sewage Projects, Tallahassee, FL, 1996.

Friedman, D. 2007. *The Septic Systems Information Website - Inspecting, Testing, Designing, & Maintaining Residential Septic Systems.* http://www.inspect-ny.com/septbook.htm

Gajurel, D.R., O. Benn, Z. L, J. Behrendt, and R. Otterpohl. 2003. Pre-treatment of Domestic Wastewater with Pre-composting Tanks: Evaluation of Existing Systems. *Water Sci. Technol.*, v. 48, no. 11-12, pp. 133-138.

Gordon, D. M., S. Bauer, and J.R. Johnson. 2002 .The Genetic Structure of Escherichia coli Populations in Primary and Secondary Habitats. *Microbiology*. v. 148, pp. 1513 – 1522.

Grady, C.P.L., G.T. Daigger, and H.C. Lim. 1998. *Biological Wastewater Treatment, Second Edition.* Taylor and Francis Group, CRC Press: Boca Raton, FL, 1998.

Greer, B.A. and W.C. Boyle. 1987. *Volatile organic compounds (VOCs) in small community wastewater disposal systems using soil absorption*. Proceedings of the Fifth National Symposium on Individual and Small Community Sewage Systems, Amer. Soc. Ag. Engineers, Chicago, IL, December 14 - 15, 1987.

Griffin, Jr., D.M., X. Yan, H. Xiang, C.A. Fletcher, and R. Crawford. 2002. *Long Term Performance of a Septic Tank – Rock Filter System.* Proceedings of the Water Environment Federation Annual Technical Exhibition and Conference (WEFTEC), Chicago, IL, September 29 – October 2, 2002.

Grimes, B.H. 2007 *Microbial Septic Tank Sampling : Investigations When a Tank Appears to be "Dead".* Proceedings of the Twenty Fourth Annual Onsite Wastewater Conference, Raleigh, NC, April 24 - 26, 2007.

Gross, M.A. 1987. *Assessment of the Effects of Household Chemicals Upon Individual Septic Tank Performances.* Arkansas Water Resources Research Center. http://www.uark.edu/depts/awrc/Publications/PUB0131.pdf

Groves, T.W., F. Bowers, E.J. Corriveau, J. Higgins, J. Heltshe, and M.T. Hoover. 2005. *Variability and Reliability of Test Center and Field Data: Definition of Proven Technology From a Regulatory Viewpoint.* Project No. WU-HT-03-35. Prepared for the National Decentralized Water Resources Capacity Development Project, Washington University, St. Louis, MO, by the New England Interstate Water Pollution Control Commission, Lowell, MA, 2005.

Gunn, I. 1998. *Joint Australia / New Zealand Performance Standard for On-site Wastewater Management.* Proceedings of the Eighth National Symposium on Individual and Small Community Sewage Systems, Amer. Soc. Ag. Engineers, Orlando, FL, March 8 - 10, 1998.

Hake, J., A. Cohn, D. Gray, and M. Ramanathan. 2006. *Get More Juice Out of That Lemon: Ebud's Search for Energy in Wastewater.* Proceedings of the Water Environment Federation Annual Technical Exhibition and Conference (WEFTEC 06), Dallas, TX, October 2006.

Haldeman, D., T. Cleveland, and J. Johnson. 2004. *Septic Tank Waste Strength and Sampling Onsite Systems: The Nuts and Bolts.* Proceedings of the National Onsite Water Recycling Association, 13th Annual conference, November 7 - 11, Albuquerque, NM, 2004.

Haseltine, T.R. 1950. *Addition of Garbage to Sewage.* Water and Sewage Works, Annual Compilation. Springer Publishing: New York, November 1950, pp. 467-472.

Hawk, G. 1982. *An Investigation of Factors Determining Grease and Oil Removal Efficiencies of Interceptors.* Report, Wastewater Treatment Division, Industrial Waste Section, City of Austin, Texas, 1982.

Hazen, A. 1904. *On Sedimentation.* Trans. ASCE. Paper No. 980, Proceedings of Annual Meeting of June 1, 1904.

Hellstrom, D. and L. Jonsson. 2003. Evaluation of Small Wastewater Treatment Systems. *Water Sci. Technol.*, v. 48, no. 11-12, pp. 61-68.

Hinkle, S.R., R.J. Weick, J.M. Johnson, J.D. Cahill, S.G. Smith, and B.J. Rich. 2005. *Organic Wastewater Compounds, Pharmaceuticals, and Coliphage in Groundwater Receiving Discharge From Onsite Wastewater Treatment Systems Near La Pine, Oregon: Occurrence and Implications for Transport.* Project No. WU-HT-03-05. Prepared for the National Decentralized Water Resources Capacity Development Project, Washington University, St. Louis, MO, by Oregon Department of Environmental Quality, Portland, OR, 2005.

Hoover, M.T., D.M. Sievers, and D.M. Gustafson. 1998. *Performance Standards for On-site Wastewater Treatment Systems.* Proceedings of Eighth National Symposium on Individual and Small Community Sewage Systems, ASAE, American Society of Agricultural and Biological Engineers, March 8 - 10, Orlando, FL, 1998.

Ingersoll, A.C., J.E. McKee, and N.H. Brooks. 1955. *Fundamental Concepts of Rectangular Settling Tanks.* Proceedings, American Society of Civil Engineers, Sanitary Engineering Division, v. 81, pp. 590-591.

Jantrania, A.R., W.A. Sack, and V. Earp. 1994. *Evaluation of Additives for Improving Septic Tank Operation under Stress Conditions.* Proceedings of the Seventh International Symposium on Individual and Small Community Sewage Systems, Amer. Soc. Ag. Engineers, Atlanta, GA, December 11 - 13, 1994.

Jantrania, A.R. and M.A. Gross. 2006. *Advanced Onsite Wastewater Systems Technologies.* Taylor and Francis Group, CRC Press: Boca Raton, FL, 2006.

Jones, E.E. 1974. *Domestic Water Use in Individual Homes and Hydraulic Loading of and Discharge from Septic Tanks.* Proceedings of the National Home Sewage Disposal Symposium; 1974; Amer. Soc. Ag. Engineers, University of Washington, Publication Proc-175.

Jones, E.E. 1978. *Septic Tank – Configuration Versus Performance.* Proceedings of the Second Northwest On-Site Wastewater Treatment Short Course and Equipment Exhibition, 1978.

Jowett, C. and R. Lay. 2006. *Comparative Performance of Closed-conduit, Laminar Flow Septic Tanks.*, Waterloo Biofilter Systems, Rockwood, ON, Canada. Submitted to *Small Flows Quarterly*, 2006.

Joy, D.M., T.R. King, S. Maunoir, and H. Philip. 2004(a). *Field Testing of the Eparco Compact Filter Wastewater System.* Proceedings of the Tenth National Symposium on Individual and Small Community Sewage Systems, Amer. Soc. Ag. Engineers, Sacramento, CA, March 21 - 24, 2004.

Joy, D.M., T.R. King, and A. Howes. 2004(b). *Temperature Profiles in a Granular Compact Filter and System Performance under Cold-weather Conditions.* Proceedings of the National Onsite Water Recycling Association's 13th Annual Conference, Albuquerque, NM, November 7 - 11, 2004.

Killam Associates. 1995. *Funeral Home Wastestream Audit Report.* Report. National Funeral Directors Association, Brookfield, WI, 1995.

Kommalapati, R.R. and R. Johnson. 2001. *A Literature Review on the Evaluation of Design Parameters for Modern Grease Traps and High Strength Wastes.* Texas Onsite Wastewater Treatment Research Council. http://www.towtrc.state.tx.us/ ResearchProjects/ LitReviewOnTheEval.pdf

Koottatep, T., A. Morel, W. Sri-Anant, and R. Schertenleib 2004. *Potential of the Anaerobic Baffled Reactor as Decentralised Wastewater Treatment System in the Tropics.* Proceedings of the 1st International Conference on Onsite Wastewater Treatment & Recycling, Perth, Australia, February 11 - 13, 2004.

Laak, R., J.J. Kolega, B.J. Cosenza, M.S. Weinberg, and J.H.T. Winneberger. 1975. *Feasibility Studies on Utilizing Sodium Bicarbonate with Septic Tank Systems.* Proceedings of the National Home Sewage Disposal Symposium. Amer. Soc. Ag. Engineers, Publication Proc-175. pp 202-209.

Laak, R. 1980. Multichamber Septic Tanks. *J. of the Environmental Engineering Division,* v. 106, No. EE3.

LaKind and Bouwer. 2003. *Investigation of the Removal of Formaldehyde and Phenol by Funeral Home Septic Systems.* Report. National Funeral Directors Association, Brookfield, WI, 2003.

Lens, P., G. Zeeman, and G. Lettinga. 2001. *Decentralised Sanitation and Reuse.* IWA Publishing: London, March 2001, ISBN 1900222477.

Lesikar, B.J., O.A. Garza, R.A. Persyn, M.T. Anderson, and A.L. Kenimer. 2004. *Food-service Establishments Wastewater Characterization*. Proceedings of the Tenth National Symposium on Individual and Small Community Sewage Systems, Amer. Soc. Ag. Engineers, Sacramento, CA, March 21 - 24, 2004.

Lesikar, B.J., O.A. Garza, R.A. Persyn, A.L. Kenimer, and M.T. Anderson . 2005. *The Strength of Wastewater as Impacted by Restaurant Management Practices*. Proceedings of the National Onsite Water Recycling Association, 14th Annual Conference, Cleveland, OH, October 10 - 13, 2005.

Lettinga, G., A.F.M. van Velsen, S.W. Hobma, W.J. de Zeeuw, and A. Klapwijk. 1980. Use of the Upflow Sludge Blanket (USB) Reactor Concept for Biological Wastewater Treatment. *Biotechnology and Bioengineering*, v. 22, pp. 699-734.

Lombardo, P. 2006. *Phosphorus geochemistry in septic tanks, soil absorption systems and groundwater*. Lombardo Associates, Inc., Newton, MA.

Long, T. 1997. *Septic Additives in Washington State.* Proceedings of the Ninth Northwest On-site Wastewater Treatment Short Course and Equipment Exhibition, 1997.

Loudon, T.L., T.R. Bounds, J.C. Converse, T. Konsler, and C. Rock. 2005. *Septic Tanks Text* in (D.L. Lindbo and N.E. Deal eds.) *Model Decentralized Wastewater Practitioner Curriculum.* National Decentralized Water Resources Capacity Development Project. North Carolina State University, Raleigh, NC.

Lowe, K.S., N.K. Rothe, J.M.B. Tomaras, K. DeJong, M.B. Tucholke, J. Drewes, J.E. McCray, and J. Munakata-Marr. 2007. *Influent characteristics of the modern waste stream from single sources: literature review.* Draft Final Report, Water Environment Research Foundation, Project No. 04-DEC-01.

Ludwig, H.F. 1949. Septic Tank Performance Under Surge Flow Conditions. *Water and Sewage Works*. March 1949. p. 122.

Ludwig, H.F. 1950. Septic Tanks: Design and Performance. *Sewage and Industrial Wastes – the Journal of the Federation of Sewage Works Association*, January 1950, Vol. 22, No. I.

Lueck, G.L. and B.H. Shaw. 1994. *On-site Disposal of Motor Vehicle Waste Fluids : an Evaluation of Current Methods and Alternatives.* Proceedings of the Seventh International Symposium on Individual and Small Community Sewage Systems, Amer. Soc. Ag. Engineers, Atlanta, GA, December 11 - 13, 1994.

Luostarinen, S. and J. Rintala. 2006. Anaerobic On-site Black Water and Kitchen Waste Treatment Using UASB-Septic Tanks at Low Temperatures. *Water Sci. Technol.*, v. 54, no. 2, pp 143-149.

Mancl, K.M. 1984. Estimating Septic Tank pumping Frequency. *J. Environ. Eng.*, v. 110, no. 1.

Mayer, P.W., W.B. DeOreo, E.M. Opitz, J.C. Kiefer, W.Y. Davis, B. Dziegielewski, and J.O. Nelson. 1999. *Residential End Uses of Water*. American Water Works Research Foundation. http://www.awwa.org/bookstore/product.cfm?id=90781

McCorquodale, J.A., A. Griborio, and I. Georgiou. 2006. *Application of a CFD Model to Improve the Performance of Rectangular Clarifiers.* Proceedings of the Water Environment Federation Annual Technical Exhibition and Conference (WEFTEC 06), pp. 310-320.

Matassa, M., C.L. McEntyre, and J.T. Watson. 2005. *Tennessee Valley Marina and Campground Wastewater characterization Screening Study.* Water Environment Federation Annual Technical Exhibition and Conference (WEFTEC 05), 2005.

Nair, G., A.K. Pati, and M.L. Naik . 1999. Temporal Organization in Population Density of Protozoans in Septic Tank Sewage. *Current Science*, v. 76, no. 8, 1128-1134.

Nguyen, V.A., P.T. Nga, T.H. Nhue, and A. Morel, K. 2007.Tonderski. *Improved Septic Tank with Constructed Wetland, a Promising Decentralized Wastewater Treatment Alternative in Vietnam.* Proceedings of the NOWRA 2007 Annual Conference and Exhibit, National Onsite Wastewater Recycling Association, 2007.

Ollivant. M. 1993. *Sanitary Sewer Rehabilitation, City of Montesano, WA.* Parametrix, Inc. Consultants in Engineering and Environmental Sciences.

Otis, J.J., W.C. Boyle, and D.K. Sauer. 1974. *The Performance of Household Wastewater Treatment Units under Field Conditions.* Proceedings of the National Home Sewage Disposal Symposium, Amer. Soc. Ag. Engineers, Publication Proc-175.

Otis, R.J. and W.C. Boyle. 1976. *Performance of Single Household Treatment Units. ASCE Journal of the Environmental Engineering Division.* Vol. 102, No. EE1.

Otis, R.J., D.L. Anderson, and R.A. Apfel. 1993. *Onsite Sewage Disposal System Research In Florida An Evaluation of Current Onsite Sewage Disposal System (OSDS) Practices in Florida.* State of Florida Department of Health and Rehabilitative Services Report. Tallahassee, FL, 1993.

Pearson, F. 1986. Septic Tanks Performance. *Small Flows.* May 1986.

Pearson, F., D. Jenkins, H. McLean, and S.A. Klein. 1980. *Recreation Vehicle Waste Disposal in Roadside Rest Septic Tank Systems.* Report. Sanitary Engineering Research Laboratory, University of California, Berkeley, CA.

Pearson, F., H. McLean and S.A. Klein. 1991. Pilot-Scale Septic Tank Treatment of Preservative-Laden Waste. *J. Water Pollut. Control Fed.*, Vol. 63, No. 7.

Pettit, T. and H. Linn. 1987. *A Guide to Safety in Confined Spaces.* National Institute for Occupational Safety and Health, U.S. Department of Health and Human Services.

Peeples, J.A. and K.M. Mancl. 1998. Laboratory Scale Septic Tanks. *Ohio Journal of Science.* v. 98, no. 4/5, pp. 75-79.

Philip, H., A. Rambaud, and J. Bontoux. 1987. Role and Performance of Septic Tanks in Individual Sanitation: Effect of Biological Activators. *Water Sci. Technol.*, v. 19, no. 7.

Philip, H., S. Maunoir, A. Rambaud, and S. Philippi. 1993. *Septic Tank Sludges: Accumulation Rate Biochemical Characteristics.* Proceedings of the Second International Specialized Conference on Design and Operation of Small Wastewater Treatment Plants, Norway.

Pradhan, S., M.T. Hoover, G.H. Clark, M. Gumpertz, A.G. Wollum, C. Cobb, and J. Strock. 2006. Septic Tank Additive Impacts on Microbial Populations. Manuscript submitted to *J. Env. Health*, October 25, 2006.

Rich, B.J., 2006. Unpublished transcript, pending publication.

Richardson, J.E., D.C. McAvoy, B. Low, and R. Edison. 2002. *Computational Fluid Dynamics (CFD) Modeling Used to Assess the Fate of Flushable Solid consumer Products in Septic Tanks.*

Proceedings of the Water Environment Federation Annual Technical Exhibition and Conference (WEFTEC 02), LOCATION, DATES, 2002.

Rock, C.A. and J.A. Boyer. 1995. *NPCA Septic Tank Project 1990-1995.* National Precast Concrete Association, University of Maine, Orono, Maine.

Russell, J.M. and M. Bigelow. 2002. *Advances in Separation Tank Monitoring Solutions Shift the Focus of On-Site Wastewater Management Programs from Cleanup and Crisis Management to Preventative Maintenance.* Proceedings of the NOWRA 2002 Annual Conference and Exhibit, Protecting Water Quality On Site, National Onsite Wastewater Recycling Association, 2002.

Sauer, P.A. and E.J. Tyler. 1992. *Motor Vehicle Waste Fluid Impacts on Septic Tank / Wastewater Infiltration Systems.* Report. Small Scale Waste Management Project, University of Wisconsin, Prepared for the 7th Northwest On-Site Wastewater Treatment Short Course and Equipment Exhibition, September 14-15, 1992.

Sauer, P.A. and E.J. Tyler. 1996. Heavy Metal and Volatile Organic Chemical Removal and Treatment in On-site Wastewater Systems. *Water, Air, and Soil Pollution*, v. 89, pp. 221-232.

Schmidt, H.E. 1976. *The "Suburbanaer" Interceptor Tank Solids Accumulation Survey.* Unpublished.

Scow, K. 1994. *The Efficacy and Environmental Impact of Biological Additives to Septic Systems.* University of California Press: Davis, CA, 1994.

Seabloom, R.W., T.R. Bounds, and T.L. Loudon. 2005. Septic Tanks Text. *in* (M.A. Gross and N.E. Deal, eds.) *University Curriculum Development for Decentralized Wastewater Management.* National Decentralized Water Resources Capacity Development Project. University of Arkansas, Fayetteville, AR.

Sherman, K.M. and D.L. Anderson. 1991. *An Evaluation of Volatile Organic Compounds and Conventional Parameters from On-Site Sewage Disposal Systems in Florida.* Proceedings of the Sixth National Symposium on Individual and Small Community Sewage Systems, Amer. Soc. Ag. Engineers, December 16 - 17, 1991.

Siegrist, R.L., M. Witt, and W.C. Boyle. 1976. Characteristics of Rural Household Wastewater. *ASCE Journal of the Environmental Engineering Division* 1976, Vol. 102, No. EE3, pp. 533-548.

Siegriest, R.L. 1977(a). *Segregation and Separate Treatment of Black and Grey Household Wastewaters to Facilitate Onsite Surface Disposal.* Report. University of Wisconsin Press: Madison, WI, 1977.

Siegriest, R.L. 1977(b). *Waste Segregation to Facilitate Onsite Wastewater Disposal Alternatives.* Proceedings of the Second National Home Sewage Treatment Symposium, Amer. Soc. Ag. Engineers, 1977; Publication 5-77.

Siegrist, R.L. and W.C. Boyle. 1981. *Onsite Reclamation of Residential Greywater.* Proceedings of the Third National Symposium on Individual and Small Community Sewage Systems, Amer. Soc. Ag. Engineers, 1981.

Siegrist, R.L., D.L. Anderson, and J.C. Converse. 1984(a). *Commercial Wastewater On-site Treatment and Disposal.* Proceedings of the Fourth National Symposium on Individual and

Small Community Sewage Systems, ASAE Publication 07-85. Amer. Soc. Ag. Engineers, St. Joseph, MI. 1984. pp. 210-219.

Siegrist, R.L., D.L. Anderson, and J.C. Converse. 1984(b). *Onsite Treatment and Disposal of Restaurant Wastewater.* Small Scale Waste Management Project, University of Wisconsin - Madison. April 1984. 38 pp.

Simmons, J.D., J.O. Newman, C.W. Rose, and E.E. Jones. 1981. *Small-diameter, Variable-grade, Gravity Sewers for Septic Tank Effluent.* Proceedings of the Third National Symposium on Individual and Small Community Sewage Systems, Amer. Soc. Ag. Engineers, 1981.

Tchobanoglous, G., F.L. Burton, and H.D. Stensel. 2003. *Wastewater Engineering (Treatment Disposal Reuse) / Metcalf & Eddy, Inc.,* 4th Edition, McGraw-Hill Book Company: New York. ISBN 0-07-041878-0.

Troyan, J.J., G.J. Sewards, and R.J. Fimmel. 1984. *Comprehensive Research on Septic Tank-Soil Absorption Systems in Perth, Australia.* Proceedings of the Fourth National Symposium on Individual and Small Community Sewage Systems, Amer. Soc. Ag. Engineers, December 10 - 11, 1984.

Tyler, E.J., R.B. Corey, and M.U. Olotu. 1978. *Potential Effects of Water Softener Use on Septic Tank – Soil Absorption On-site Waste Water Systems.* University of Wisconsin Press: Madison, WI, 1978.

US EPA. 1987. Guidance Manual for Preventing Interference at POTWs. 833B87201. September 1987.

Vaishnav, D.D. and J.W. McCabe. 1996. Development of an Anaerobic Sludge Respiration Inhibition Test and Its Use to Assess Septic Tank Safety of Consumer Products. *Environ. Toxicol. Chem.* 1996, v. 15, no. 9.

Viraraghavan, T. and S. Hashem. 1986. Trace Organics in Septic Tank Effluent. *Water, Air Soil Pollut.* 1986, v. 28, no. 34, p. 299-308.

Wallace, S.D. and C.D. English. 2003. *Cold weather insulation of septic tanks.* Proceedings of the National Onsite Water Recycling Association, 12th Annual Conference, Nashville, TN, November 3 - 6, 2003.

Warshall, P. 1979. *Septic Tank Practices.* Anchor Press/Doubleday: Garden City, NY, 1979.

Weibel, S.R., T.W. Bendixen, and J.B. Coulter. 1949. *Studies on Household Sewage Disposal Systems, Part I.* US Public Health Service, Cincinnati, OH.

Weibel, S.R., T.W. Bendixen, and J.B. Coulter. 1954. *Studies on Household Sewage Disposal Systems, Part II.* US Public Health Service, Cincinnati, OH.

Weibel, S.R., T.W. Bendixen, and J.B. Coulter. 1955. *Studies on Household Sewage Disposal Systems, Part III.* US Public Health Service, Cincinnati, OH.

Winneberger, J.H.T. 1984. *Septic-tank Systems A Consultant's Toolkit. Volume II – The Septic Tank.* The Butterworth Group, Ann Arbor, MI, 1984.

Wolfe. 1999. *World of Water 2000 – The Past, Present and Future.* Water World/Water and Wastewater International Supplement to PennWell Magazines, Tulsa, OK.

Wong, M., R. Ngie-Hing, and R. Lowhorn. 1998. *Total Suspended Solids and Oil & Grease Removal by a Grease-trap Effluent.* Proceedings of the National Onsite Water Recycling Association, 7th Annual Conference, Ft. Mitchell, KY, 1998.

Yost, S. and S. Lingireddy. 1997. *Hydraulic Analysis of Vaulted Pumps in Septic Tanks.* Proceedings of the National Onsite Water Recycling Association, 6th Annual Conference, LOCATION, DATES.

Zanini, L., W. D. Robertson, C.J. Ptacek, S.L. Schiff, and T. Mayer. 1998. Phosphorus Characterization in Sediments Impacted by Septic Effluent at Four Sites in Central Canada. *J. Contam. Hydrol.* 1998, v. 33, pp. 405-429.

WASTEWATER UTILITY

Alabama
Montgomery Water Works & Sanitary Sewer Board

Alaska
Anchorage Water & Wastewater Utility

Arizona
Glendale, City of, Utilities Department
Mesa, City of
Peoria, City of
Phoenix Water Services Dept.
Pima County Wastewater Management
Safford, City of

Arkansas
Little Rock Wastewater Utility

California
Central Contra Costa Sanitary District
Corona, City of
Crestline Sanitation District
Delta Diablo Sanitation District
Dublin San Ramon Services District
East Bay Dischargers Authority
East Bay Municipal Utility District
Eastern Municipal Water District
El Dorado Irrigation District
Fairfield-Suisun Sewer District
Fresno Department of Public Utilities
Inland Empire Utilities Agency
Irvine Ranch Water District
Las Virgenes Municipal Water District
Livermore, City of
Los Angeles, City of
Los Angeles County, Sanitation Districts of
Napa Sanitation District
Orange County Sanitation District
Palo Alto, City of
Riverside, City of
Sacramento Regional County Sanitation District
San Diego Metropolitan Wastewater Department, City of
San Francisco, City & County of
San Jose, City of
Santa Barbara, City of
Santa Cruz, City of
Santa Rosa, City of
South Bayside System Authority
South Coast Water District
South Orange County Wastewater Authority
Stege Sanitary District
Sunnyvale, City of
Union Sanitary District
West Valley Sanitation District

Colorado
Aurora, City of
Boulder, City of
Greeley, City of
Littleton/Englewood Water Pollution Control Plant
Metro Wastewater Reclamation District, Denver

Connecticut
Greater New Haven WPCA
Stamford, City of

District of Columbia
District of Columbia Water & Sewer Authority

Florida
Broward, County of
Fort Lauderdale, City of
Miami-Dade Water & Sewer Authority
Orange County Utilities Department
Reedy Creek Improvement District
Seminole County Environmental Services
St. Petersburg, City of
Tallahassee, City of
Tampa, City of
Toho Water Authority
West Palm Beach, City of

Georgia
Atlanta Department of Watershed Management
Augusta, City of
Clayton County Water Authority
Cobb County Water System
Columbus Water Works
Fulton County
Gwinnett County Department of Public Utilities
Savannah, City of

Hawaii
Honolulu, City & County of

Idaho
Boise, City of

Illinois
American Bottoms Wastewater Treatment Plant
Greater Peoria Sanitary District
Kankakee River Metropolitan Agency
Metropolitan Water Reclamation District of Greater Chicago
Wheaton Sanitary District

Iowa
Ames, City of
Cedar Rapids Wastewater Facility
Des Moines, City of
Iowa City

Kansas
Johnson County Unified Wastewater Districts
Unified Government of Wyandotte County/ Kansas City, City of

Kentucky
Louisville & Jefferson County Metropolitan Sewer District
Sanitation District No. 1

Louisiana
Sewerage & Water Board of New Orleans

Maine
Bangor, City of
Portland Water District

Maryland
Anne Arundel County Bureau of Utility Operations
Howard County Department of Public Works
Washington Suburban Sanitary Commission

Massachusetts
Boston Water & Sewer Commission
Massachusetts Water Resources Authority (MWRA)
Upper Blackstone Water Pollution Abatement District

Michigan
Ann Arbor, City of
Detroit, City of
Holland Board of Public Works
Saginaw, City of
Wayne County Department of Environment
Wyoming, City of

Minnesota
Rochester, City of
Western Lake Superior Sanitary District

Missouri
Independence, City of
Kansas City Missouri Water Services Department
Little Blue Valley Sewer District
Metropolitan St. Louis Sewer District

Nebraska
Lincoln Wastewater System

Nevada
Henderson, City of
Las Vegas, City of
Reno, City of

New Jersey
Bergen County Utilities Authority
Ocean County Utilities Authority
Passaic Valley Sewerage Commissioners

New York
New York City Department of Environmental Protection

North Carolina
Charlotte/Mecklenburg Utilities
Durham, City of
Metropolitan Sewerage District of Buncombe County
Orange Water & Sewer Authority

Ohio
Akron, City of
Butler County Department of Environmental Services
Columbus, City of
Metropolitan Sewer District of Greater Cincinnati
Northeast Ohio Regional Sewer District
Summit, County of

Oklahoma
Oklahoma City Water & Wastewater Utility Department
Tulsa, City of

Oregon
Albany, City of
Clean Water Services
Eugene, City of
Gresham, City of
Portland, City of Bureau of Environmental Services
Water Environment Services

Pennsylvania
Hemlock Municipal Sewer Cooperative (HMSC)
Philadelphia, City of
University Area Joint Authority

South Carolina
Charleston Water System
Mount Pleasant Waterworks & Sewer Commission
Spartanburg Water

Tennessee
Cleveland, City of
Knoxville Utilities Board
Murfreesboro Water & Sewer Department
Nashville Metro Water Services

Texas
Austin, City of
Dallas Water Utilities
Denton, City of
El Paso Water Utilities
Fort Worth, City of
Houston, City of
San Antonio Water System
Trinity River Authority

Utah
Salt Lake City Corporation

Virginia
Alexandria Sanitation Authority

WERF SUBSCRIBERS

Arlington, County of
Fairfax County
Hampton Roads Sanitation District
Hanover, County of
Henrico, County of
Hopewell Regional Wastewater Treatment Facility
Loudoun County Sanitation Authority
Lynchburg Regional WWTP
Prince William County Service Authority
Richmond, City of
Rivanna Water & Sewer Authority

Washington
Everett, City of
King County Department of Natural Resources
Seattle Public Utilities
Sunnyside, Port of
Yakima, City of

Wisconsin
Green Bay Metro Sewerage District
Kenosha Water Utility
Madison Metropolitan Sewerage District
Milwaukee Metropolitan Sewerage District
Racine, City of
Sheboygan Regional Wastewater Treatment
Wausau Water Works

Australia
ACTEW (Ecowise)
South Australian Water Corporation
Sydney Water Corporation
Water Corporation of Western Australia

Canada
Lethbridge, City of
Regina, City of, Saskatchewan
Toronto, City of, Ontario
Winnipeg, City of, Manitoba

New Zealand
Watercare Services Limited

United Kingdom
United Utilities North West (UUNW)

STORMWATER UTILITY

California
Fresno Metropolitan Flood Control District
Los Angeles, City of, Department of Public Works
Monterey, City of
Sacramento, County of
San Francisco, City & County of
Santa Rosa, City of
Sunnyvale, City of

Colorado
Aurora, City of
Boulder, City of

Georgia
Griffin, City of

Iowa
Cedar Rapids Wastewater Facility
Des Moines, City of

Kansas
Overland Park, City of

Kentucky
Louisville & Jefferson County Metropolitan Sewer District

Maine
Portland Water District

North Carolina
Charlotte, City of, Stormwater Services

Pennsylvania
Philadelphia, City of

Tennessee
Chattanooga Stormwater Management

Texas
Harris County Flood Control District, Texas

Washington
Bellevue Utilities Department
Seattle Public Utilities

STATE

Arkansas Department of Environmental Quality
Connecticut Department of Environmental Protection
Kansas Department of Health & Environment
Kentucky Department of Environmental Protection
New England Interstate Water Pollution Control Commission (NEIWPCC)
Ohio River Valley Sanitation Commission
Urban Drainage & Flood Control District, CO

CORPORATE

ADS Environmental Services
Advanced Data Mining International
Alan Plummer & Associates
Alpine Technology Inc.
Aqua-Aerobic Systems Inc.
Aquateam–Norwegian Water Technology Centre A/S
ARCADIS
Associated Engineering
Black & Veatch
Blue Water Technologies, Inc.
Boyle Engineering Corporation
Brown & Caldwell
Burgess & Niple, Ltd.
Burns & McDonnell
CABE Associates Inc.
The Cadmus Group
Camp Dresser & McKee Inc.
Carollo Engineers Inc.
Carpenter Environmental Associates Inc.
CDS Technologies Inc.
CET Engineering Services
Chemtrac Systems Inc.
CH2M HILL
CRA Infrastructure & Engineering
CONTECH Stormwater Solutions
D&B/Guarino Engineers, LLC
Damon S. Williams Associates, LLC
Earth Tech Inc.
Ecovation
EMA Inc.
Environmental Operating Solutions, Inc.
Environ/The ADVENT Group, Inc.
Fay, Spofford, & Thorndike Inc.
Freese & Nichols Inc.
ftn Associates Inc.
Gannett Fleming Inc.
Garden & Associates, Ltd.
Geosyntec Consultants
GHD
Golder Associates Ltd.
Greeley and Hansen LLC
Hazen & Sawyer, P.C.
HDR Engineering Inc.
HNTB Corporation
Hydromantis Inc.
HydroQual Inc.
Infilco Degremont Inc.
Jacobson Satchell Consultants, Inc.
Jacques Whitford NAWE, Inc.
Jason Consultants LLC Inc.
Jordan, Jones, & Goulding Inc.
KCI Technologies Inc.
Kelly & Weaver, P.C.
Kennedy/Jenks Consultants
KMK Consultants
Komline Sanderson Engineering Corporation
Larry Walker Associates
Limno-Tech Inc.
The Low Impact Development Center Inc.
Malcolm Pirnie Inc.
Material Matters
McKim & Creed
Metcalf & Eddy Inc.
Monteco Corporation
MPR Engineering Corporation, Inc.
MWH
NewFields Water Resources, LLC
O'Brien & Gere Engineers Inc.
Odor & Corrosion Technology Consultants Inc.
Original Engineering Consultants, Ltd.
Oscar Larson & Associates
Parametrix Inc.
Parsons
Post, Buckley, Schuh & Jernigan
Praxair, Inc.
Ring Industrial Group
RMC Water & Environment
Ross & Associates Ltd.
Rothberg, Tamburini & Windsor, Inc.
SAIC
Savin Engineers
Siemens Water Technologies
Stantec Consulting Inc.
Stearns & Wheler, LLC
Stone Environmental Inc.
Stratus Consulting Inc.
Synagro Technologies Inc.
Tetra Tech Inc.
Trojan Technologies Inc.
Trussell Technologies, Inc.
URS Corporation
Wade-Trim Inc.
Westin Engineering Inc.
Weston Solutions Inc.
Woodard & Curran
Zenon Environmental Inc./GE Water Process Technology
Zoeller Pump Company

INDUSTRY

American Electric Power
American Water
ChevronTexaco Energy Research & Technology Company
The Coca-Cola Company
Dow Chemical Company
DuPont Company
Eastman Chemical Company
Eli Lilly & Company
Merck & Company Inc.
Premier Chemicals LLC
Procter & Gamble Company
Thames Water Utilities
Severn Trent Services Inc.
Suez Environnment
United Water Services LLC

Note: List as of 10/31/07

WERF Board of Directors

WERF Research Council

IWA Publishing's authorised EU representative for General Product Safety Regulations is Diane D'Arras, 15 rue Duret, 75116 Paris, France, e-mail: safety@iwap.co.uk.

Printed and bound by CPI Group (UK) Ltd, Croydon, CR0 4YY

12/05/2026

02108865-0011